Joint FAO/WHO Food Standards

CODEX ALIMENTARIUS COMMISSION

PROCEDURAL MANUAL
Tenth edition

FOOD AND AGRICULTURE ORGANIZATION OF THE UNITED NATIONS
WORLD HEALTH ORGANIZATION
Rome, 1997

> The designations employed and the presentation of material in this publication do not imply the expression of any opinion whatsoever on the part of the Food and Agriculture Organization of the United Nations or of the World Health Organization concerning the legal status of any country, territory, city or area or of its authorities, or concerning the delimitation of its frontiers or boundaries.

M-83
ISBN 92-5-104014-1

All rights reserved. No part of this publication may be reproduced, stored in a retrieval system or transmitted in any form or by any means, electronic, mechanical, photocopying or otherwise, without the prior permission of the copyright owner. Applications for such permission, with a statement of the purpose and extent of the reproduction, should be addressed to the Director, Information Division, Food and Agriculture Organization of the United Nations, Viale delle Terme di Caracalla, 00100 Rome, Italy.

© **FAO and WHO 1997**

CONTENTS

CONTENTS iii

INTRODUCTION 1

STATUTES OF THE CODEX ALIMENTARIUS COMMISSION 4

RULES OF PROCEDURE OF THE CODEX ALIMENTARIUS
COMMISSION 7

PROCEDURES FOR THE ELABORATION OF CODEX STANDARDS
AND RELATED TEXTS 18

GENERAL PRINCIPLES OF THE CODEX ALIMENTARIUS 29

GUIDELINES FOR THE ACCEPTANCE PROCEDURE FOR CODEX
STANDARDS 35

DEFINITIONS FOR THE PURPOSES OF THE CODEX
ALIMENTARIUS 42

GUIDELINES FOR CODEX COMMITTEES 48

GUIDELINES FOR THE INCLUSION OF SPECIFIC PROVISIONS IN
CODEX STANDARDS AND RELATED TEXTS 56

UNIFORM SYSTEM OF REFERENCES FOR CODEX DOCUMENTS . 67

FORMAT FOR CODEX COMMODITY STANDARDS INCLUDING
STANDARDS ELABORATED UNDER THE CODE OF PRINCIPLES
CONCERNING MILK AND MILK PRODUCTS 69

CRITERIA FOR THE ESTABLISHMENT OF WORK PRIORITIES AND
FOR THE ESTABLISHMENT OF SUBSIDIARY BODIES OF THE
CODEX ALIMENTARIUS COMMISSION 73

RELATIONS BETWEEN COMMODITY COMMITTEES AND
GENERAL COMMITTEES 75

SESSIONS OF THE CODEX ALIMENTARIUS COMMISSION 84

LIST OF SUBSIDIARY BODIES OF THE CODEX ALIMENTARIUS COMMISSION 85

MEMBERS OF THE CODEX ALIMENTARIUS COMMISSION 113

LIST OF CODEX CONTACT POINTS 116

APPENDIX: GENERAL DECISIONS OF THE COMMISSION 146

INDEX 148

INTRODUCTION

The Procedural Manual of the Codex Alimentarius Commission is intended to help Member governments participate effectively in the work of the Joint FAO/WHO Food Standards Programme. The Manual is particularly useful for national delegations attending Codex meetings and for international organizations attending as observers. It will also be useful for Member Governments which wish to participate in Codex work by correspondence.

Section I sets out the Commission's basic Rules of Procedure and the other internal procedures necessary to achieve the Commission's objectives. These include the procedures for the elaboration of Codex Standards and related texts, general principles and guidelines for the acceptance of Codex standards by governments, and some basic definitions.

Section II is devoted to guidelines for the efficient operation of Codex Committees. These Committees are organized and operated by Member Governments designated by the Commission. It describes how standards are set out in a uniform manner, describes a uniform reference system for Codex documents and working papers, and provides a number of general principles for formulating key sections of Codex standards.

Section III lists the Commission's subsidiary bodies with their Terms of Reference. It also gives the Membership of the Commission (157 Member countries in July 1997) together with the addresses of Codex Contact Points.

This Tenth Edition of the Procedural Manual was prepared by the Secretariat following the Twenty-Second Session of the Codex Alimentarius Commission, Geneva 1997. Further information concerning the Codex Alimentarius Commission and its Subsidiary Bodies can be obtained from the Secretary, Codex Alimentarius Commission, Joint FAO/WHO Food Standards Programme, FAO, 00100 Rome, Italy.

SECTION I

Statutes

Rules of Procedure

Elaboration Procedures for Codex Standards

General Principles and Acceptance of Codex Standards

Definitions

CONTENTS OF THIS SECTION

The Statutes and Rules of Procedure of the Codex Alimentarius Commission were first established by FAO Conference and the World Health Assembly in 1961/62 when the Commission itself was established. The Statutes were revised in 1966. The Rules of Procedure have been amended on several occasions, the last time being in 1995. The Statutes form the legal basis of the Commission's work and provide its mandate or terms of reference. The Rules of Procedure describe the formal working procedures appropriate to an intergovernmental body.

The Procedure for the Elaboration of Codex Standards describes the way by which Codex standards are prepared and the various Steps in the process which ensure comprehensive review of draft standards by governments and other interested parties. It was comprehensively revised in 1993 to provide a uniform elaboration procedure for all Codex standards and related texts. The Procedure allows the use of a "fast-track" approach in cases where urgent action is needed.

The General Principles of the Codex Alimentarius define the Scope and the purpose of Codex Standards and the way by which governments indicate their formal acceptance of the Standards. The Guidelines on Acceptance provide additional information to Member governments on the procedures regarding acceptance.

This Section concludes with Definitions for the Purpose of the Codex Alimentarius which assist in the uniform interpretation of these texts.

STATUTES OF THE CODEX ALIMENTARIUS COMMISSION

ARTICLE 1

The Codex Alimentarius Commission shall, subject to Article 5 below, be responsible for making proposals to, and shall be consulted by, the Directors General of the Food and Agriculture Organization (FAO) and the World Health Organization (WHO) on all matters pertaining to the implementation of the Joint FAO/WHO Food Standards Programme, the purpose of which is:

(a) protecting the health of the consumers and ensuring fair practices in the food trade;

(b) promoting coordination of all food standards work undertaken by international governmental and non governmental organizations;

(c) determining priorities and initiating and guiding the preparation of draft standards through and with the aid of appropriate organizations;

(d) finalizing standards elaborated under (c) above and, after acceptance by governments, publishing them in a Codex Alimentarius either as regional or world wide standards, together with international standards already finalized by other bodies under (b) above, wherever this is practicable;

(e) amending published standards, after appropriate survey in the light of developments.

ARTICLE 2

Membership of the Commission is open to all Member Nations and Associate Members of FAO and WHO which are interested in international food standards. Membership shall comprise such of these nations as have notified the Director General of FAO or of WHO of their desire to be considered as Members.

ARTICLE 3

Any Member Nation or Associate Member of FAO or WHO which is not a Member of the Commission but has a special interest in the work of the Commission, may, upon request communicated to the Director General of FAO or WHO, as appropriate, attend sessions of the Commission and of its subsidiary bodies and ad hoc meetings as observers.

ARTICLE 4

Nations which, while not Member Nations or Associate Members of FAO or WHO, are members of the United Nations, may be invited on their request to attend meetings of the Commission as observers in accordance with the provisions of FAO and WHO relating to the grant of observer status to nations.

ARTICLE 5

The Commission shall report and make recommendations to the Conference of FAO and the appropriate body of WHO through their respective Directors General. Copies of reports, including any conclusions and recommendations, will be circulated to interested Member Nations and international organizations for their information as soon as they become available.

ARTICLE 6

The Commission shall establish an Executive Committee whose composition should ensure an adequate representation of the various geographical areas of the world to which the Members of the Commission belong. Between sessions, the Executive Committee shall act as the Executive organ of the Commission.

ARTICLE 7

The Commission may establish such other subsidiary bodies as it deems necessary for the accomplishment of its task, subject to the availability of the necessary funds.

ARTICLE 8

The Commission may adopt and amend its own Rules of Procedure which shall come into force upon approval by the Directors General of FAO and WHO, subject to such confirmation as may be prescribed by the procedures of these Organizations.

ARTICLE 9

The operating expenses of the Commission and of its subsidiary bodies, other than those for which a Member has accepted the Chair, shall be borne by the budget of the Joint FAO/WHO Food Standards Programme which shall be administered by FAO on behalf of the two Organizations in accordance with the financial regulations of FAO. The Directors General of FAO and WHO shall jointly determine the respective portion of the costs of the Programme to be borne by each Organization and prepare the corresponding annual expenditure estimates for inclusion in the Regular Budgets of the two Organizations for approval by the appropriate governing bodies.

ARTICLE 10

All expenses (including those relating to meetings, documents and interpretation) involved in preparatory work on draft standards undertaken by Members of the Commission, either independently or upon recommendation of the Commission, shall be defrayed by the government concerned. Within the approved budgetary estimates, the Commission may, however, recommend that a specified part of the costs of the preparatory work undertaken by the government on behalf of the Commission be recognized as operating expenses of the Commission.

RULES OF PROCEDURE OF THE CODEX ALIMENTARIUS COMMISSION

RULE I MEMBERSHIP

1. Membership of the Joint FAO/WHO Codex Alimentarius Commission, hereinafter referred to as "the Commission", is open to all Member Nations and Associate Members of FAO and/or WHO.

2. Membership shall comprise such eligible nations as have notified the Director General of FAO or of WHO of their desire to be considered Members of the Commission.

3. Each Member of the Commission shall communicate to the Director General of FAO or of WHO the names of its representative and where possible other members of its delegation before the opening of each session of the Commission.

RULE II OFFICERS

1. The Commission shall elect a Chairperson and three Vice Chairpersons from among the representatives, alternates and advisers (hereinafter referred to as "delegates") of the Members of the Commission; it being understood that no delegate shall be eligible without the concurrence of the head of his delegation. They shall be elected at each session and shall hold office from the end of the session at which they were elected until the end of the following regular session. The Chairperson and Vice-Chairpersons may remain in office only with the continuing endorsement of the respective Member of the Commission of which they were a delegate at the time of election. The Directors-General of FAO and WHO shall declare a position vacant when advised by the Member of the Commission that such endorsement has ceased. The Chairperson and Vice Chairpersons shall be eligible for re election but after having served two consecutive terms shall be ineligible to hold such office for the next succeeding term.

2. The Chairperson, or in his absence a Vice Chairperson, shall preside at meetings of the Commission and exercise such other function as may be required to facilitate the work of the Commission. A Vice-Chairperson acting as Chairperson shall have the same powers and duties as the Chairperson.

3. When neither the Chairperson nor the Vice-Chairperson are able to serve and, on the request of the outgoing Chairperson, during elections for the Chairperson, the Directors-General of FAO and WHO shall appoint a staff member to act as Chairperson, until either a temporary Chairperson or a new Chairperson has been elected. Any temporary Chairperson so elected shall hold

office until the Chairperson or one of the Vice-Chairpersons is able to serve again.

4. (a) The Commission may appoint a Coordinator from among the delegates of the Members of the Commission for any of the geographic locations enumerated in Rule III.1 (hereinafter referred to as 'regions') or for any group of countries specifically enumerated by the Commission (hereinafter referred to as 'groups of countries'), whenever it may find, on the basis of a proposal of a majority of the Members of the Commission which constitute the region or group, that work for the Codex Alimentarius in the countries concerned so requires.

(b) Appointments of Coordinators shall be made exclusively on the proposal of a majority of the Members of the Commission which constitute the region or group of countries concerned. Coordinators shall hold office from the end of the session of the Commission at which they were elected until not later than the end of the third succeeding regular session, the precise term being determined by the Commission in each instance. After having served two consecutive terms, the Coordinators shall be ineligible to hold such office for the next succeeding term.

(c) The functions of the Coordinators shall be:

(i) to assist and coordinate the work of the Codex Committees set up under Rule IX.1(b)(i) in their region or group of countries in the preparation of draft standards, guidelines and other recommendations for submission to the Commission;

(ii) to assist the Executive Committee and the Commission, as required, by advising them of the views of countries and recognized regional intergovernmental and non-government organizations in their respective regions on matters under discussion or of interest.

(d) Where a Coordinating Committee has been set up under Rule IX.1(b)(ii), the Coordinator of the region involved should be the Chairperson of the Committee.

(e) Where a Coordinator is not able to carry out the functions of office, the Member of the Commission of which the Coordinator was a delegate at the time of appointment shall inform the Secretary of the Commission accordingly and shall appoint an interim Coordinator until such time as the Coordinator is able to resume those functions or until the next session of the Commission.

5. The Commission may appoint one or more rapporteurs from among the delegates of the Members of the Commission.

6. The Directors-General of FAO and WHO shall be requested to appoint from the staffs of their organizations a Secretary of the Commission and such other officials, likewise responsible to them, as may be necessary to assist the officers and the Secretary in performing all duties that the work of the Commission may require.

RULE III EXECUTIVE COMMITTEE

1. The Executive Committee shall consist of the Chairperson and Vice-Chairperson of the Commission together with six further members, elected by the Commission at regular sessions from among the Members of the Commission, one each coming from the following geographic locations: Africa, Asia, Europe, Latin America and the Caribbean, North America, South-West Pacific; it being understood that not more than one delegate from any one country shall be a member of the Executive Committee. Members elected on a geographic basis shall hold office from the end of the session of the Commission at which they were elected until the end of the second succeeding regular session and shall be eligible for re election, but after having served two consecutive terms shall be ineligible to hold such office for the next succeeding term.

2. The Executive Committee shall, between sessions of the Commission, act on behalf of the Commission as its executive organ. In particular the Executive Committee may make proposals to the Commission regarding the general orientation and programme of work of the Commission, study special problems and help implement the programme as approved by the Commission. The Executive Committee may also exercise, when it shall deem it to be essential and subject to confirmation by the next session of the Commission, the Commission's powers under Rule IX.1(b)(i), Rule IX.5 insofar as it refers to bodies established under Rule IX.1(b)(i), and Rule IX.10, insofar as it refers to the designation of the Members who shall be responsible for appointing Chairpersons to subsidiary bodies established under Rule IX.1(b)(i).

3. The Chairperson and Vice Chairpersons of the Commission shall be respectively the Chairperson and Vice Chairpersons of the Executive Committee.

4. Sessions of the Executive Committee may be convened as often as necessary by the Directors-General of FAO and WHO, in consultation with the Chairperson. The Executive Committee shall normally meet immediately prior to each session of the Commission.

5. The Executive Committee shall report to the Commission.

RULE IV SESSIONS

1. The Commission shall in principle hold one regular session each year at the Headquarters of either FAO or WHO. Additional sessions shall be held as considered necessary by the Directors General of FAO and WHO after consultation with the Chairperson of the Executive Committee.

2. Sessions of the Commission shall be convened and the place of the meeting shall be determined by the Directors General of FAO and WHO after consultation where appropriate, with the authorities of the host country.

3. Notice of the date and place of each session of the Commission shall be communicated to all Members of the Commission at least two months before the session.

4. Each Member of the Commission shall have one representative, who may be accompanied by one or more alternates and advisers.

5. Meetings of the Commission shall be held in public, unless the Commission decides otherwise.

6. The majority of the Members of the Commission shall constitute a quorum for the purposes of making recommendations for amendments to the Statutes of the Commission and of adopting amendments of, or additions to, the present Rules in accordance with Rule XIII.1. For all other purposes the majority of the Members of the Commission attending the session shall constitute a quorum, provided that such a majority shall be not less than 20 percent of the total membership of the Commission, nor less than 25 Members. In addition, in the case of amendment or adoption of a proposed standard for a given region or group of countries, the quorum of the Commission shall include one third of the Members belonging to the region or group of countries concerned.

RULE V AGENDA

1. The Directors General of FAO and WHO, after consultation with the Chairperson of the Commission or with the Executive Committee, shall prepare a Provisional Agenda for each session of the Commission.

2. The first item on the Provisional Agenda shall be the adoption of the Agenda.

3. Any Member of the Commission may request the Directors-General of FAO or WHO to include specific items in the Provisional Agenda.

4. The Provisional Agenda shall be circulated by the Directors General of FAO or WHO to all Members of the Commission at least two months before the opening of the session.

5. Any Member of the Commission, and the Directors-General of FAO and WHO, may, after the dispatch of the Provisional Agenda, propose the inclusion of specific items in the Agenda with respect to matters of an urgent nature. These items shall be placed on a supplementary list, which, if time permits before the opening of the session, shall be dispatched by the Directors-General of FAO and WHO to all Members of the Commission, failing which the supplementary list shall be communicated to the Chairperson for submission to the Commission.

6. No items included in the Agenda by the governing bodies or the Directors General of FAO and WHO shall be deleted therefrom. After the Agenda has been adopted, the Commission may, by a two-thirds majority of the votes cast, amend the Agenda by the deletion, addition or modification of any other item.

7. Documents to be submitted to the Commission at any session shall be furnished by the Directors-General of FAO and WHO to all Members of the Commission, to the other eligible Nations attending the session as observers and to the non-member nations and international organizations invited as observers thereto, in principle at least two months prior to the session at which they are to be discussed.

RULE VI VOTING AND PROCEDURES

1. Subject to the provisions of paragraph 3 of this rule, each Member of the Commission shall have one vote. An alternate or adviser shall not have the right to vote except where substituting for the representative.

2. Except as otherwise provided in these rules, decisions of the Commission shall be taken by a majority of the votes cast.

3. At the request of a majority of the Members of the Commission constituting a given region or a group of countries that a standard be elaborated, the standard concerned shall be elaborated as a standard primarily intended for that region or group of countries. When a vote is taken on the elaboration, amendment or adoption of a draft standard primarily intended for a region or group of countries, only Members belonging to that region or group of countries may take part in the voting. The adoption of the standard may, however, take place only after submission of the draft text to all Members of the Commission for comments. The provisions of this paragraph shall not prejudice the elaboration or adoption of a corresponding standard with a different territorial scope.

4. Subject to the provisions of paragraph 5 of this Rule, any Member of the Commission may request a roll-call vote, in which case the vote of each Member shall be recorded.

5. Elections shall be decided by secret ballot, except that, where the number of candidates does not exceed the number of vacancies, the Chairperson may submit to the Commission that the election be decided by clear general consent. Any other matter shall be decided by secret ballot if the Commission so determines.

6. Formal proposals relating to items of the Agenda and amendments thereto shall be introduced in writing and handed to the Chairperson, who shall circulate them to representatives of Members of the Commission.

7. The provisions of Rule XII of the General Rules of FAO shall apply *mutatis mutandis* to all matters which are not specifically dealt with under Rule VI of the present Rules.

RULE VII OBSERVERS

1. Any Member Nation and any Associate Member of FAO or WHO which is not a Member of the Commission but has a special interest in the work of the Commission, may, upon request communicated to the Director-General of FAO or WHO, attend sessions of the Commission and of its subsidiary bodies as an observer. It may submit memoranda and participate without vote in the discussion.

2. Nations which, while not Member Nations or Associate Members of FAO or WHO, are Members of the United Nations, may, upon their request and subject to the provisions relating to the granting of observer status to nations adopted by the Conference of FAO and the World Health Assembly, be invited to attend in an observer capacity sessions of the Commission and of its subsidiary bodies. The status of nations invited to such sessions shall be governed by the relevant provisions adopted by the Conference of FAO.

3. Any Member of the Commission may attend as an observer the sessions of the subsidiary bodies and may submit memoranda and participate without vote in the discussions.

4. Subject to the provisions of Rule VII.5 the Directors-General of FAO or WHO may invite intergovernmental and international non-governmental organizations to attend as observers sessions of the Commission and of its subsidiary bodies.

5. Participation of international organizations in the work of the Commission, and the relations between the Commission and such organizations shall be governed by the relevant provisions of the Constitutions of FAO or WHO, as well as by the applicable regulations of FAO or WHO on relations with international organizations; such relations shall be handled by the Director-General of FAO or of WHO as appropriate.

RULE VIII RECORDS AND REPORTS

1. At each session the Commission shall approve a report embodying its views, recommendations and conclusions, including when requested a statement of minority views. Such other records for its own use as the Commission may on occasion decide shall also be maintained.

2. The report of the Commission shall be transmitted to the Directors General of FAO and WHO at the close of each session, who shall circulate it to the Members of the Commission, to other countries and to organizations that were represented at the session, for their information, and upon request to other Member Nations and Associate Members of FAO and WHO.

3. Recommendations of the Commission having policy, programme or financial implications for FAO and/or WHO shall be brought by the Directors-General to the attention of the governing bodies of FAO and/or WHO for appropriate action.

4. Subject to the provisions of the preceding paragraph, the Directors-General of FAO and WHO may request Members of the Commission to supply the Commission with information on action taken on the basis of recommendations made by the Commission.

RULE IX SUBSIDIARY BODIES

1. The Commission may establish the following types of subsidiary bodies:

   (a)[1] subsidiary bodies which it deems necessary for the accomplishment of its work in the finalization of draft standards;

   (b) subsidiary bodies in the form of:

   (i) Codex Committees for the preparation of draft standards for submission to the Commission, whether intended for world-wide use, for a given region or for a group of countries specifically enumerated by the Commission.

   (ii) Coordinating Committees for regions or groups of countries which shall exercise general coordination in the preparation of standards relating to such regions or groups of countries and such other functions as may be entrusted to them.

2. Subject to paragraph 3 below, membership in these subsidiary bodies shall consist, as may be determined by the Commission, either of such Members of the Commission as have notified the Directors-General of FAO or WHO of their desire to be considered as Members thereof, or of selected Members designated by the Commission.

3. Membership of subsidiary bodies established under Rule IX.1(b)(i) for the preparation of draft standards intended primarily for a region or group of countries, shall be open only to Members of the Commission belonging to such a region or group of countries.

4. Representatives of members of subsidiary bodies shall, insofar as possible, serve in a continuing capacity and shall be specialists active in the fields of the respective subsidiary bodies.

5. Subsidiary bodies may only be established by the Commission except where otherwise provided in these Rules. Their terms of reference and reporting procedures shall be determined by the Commission, except where otherwise provided in these Rules.

6. Sessions of subsidiary bodies shall be convened by the Directors-General of FAO and WHO:

(a) in the case of bodies established under Rule IX.1(a), in consultation with the Chairperson of the Commission;

(b) in the case of bodies established under Rule IX.1(b)(i) (Codex Committees), in consultation with the chairperson of the respective Codex Committee and also, in the case of Codex Committees for the preparation of draft standards for a given region or group of countries, with the Coordinator, if a Coordinator has been appointed for the region or group of countries concerned;

(c) in the case of bodies established under Rule IX.1(b)(ii) (Coordinating Committees), in consultation with the Chairperson of the Coordinating Committee.

7. The Directors General of FAO and WHO shall determine the place of meeting of bodies established under Rule IX.1(a) and Rule IX.1(b)(ii) after consultation, where appropriate, with the host country concerned and, in the case of bodies established under Rule IX.1(b)(ii), after consultation with the Coordinator for the region or group of countries concerned, if any, or with the Chairperson of the Coordinating Committee.

8. Notice of the date and place of each session of bodies established under Rule IX.1(a) shall be communicated to all Members of the Commission at least two months before the session.

9. The establishment of subsidiary bodies under Rule IX.1(a) and Rule IX.1(b)(ii) shall be subject to the availability of the necessary funds, as shall the establishment of subsidiary bodies under Rule IX.1(b)(i) when any of their expenses are proposed to be recognized as operating expenses within the budget of the Commission in accordance with Article 10 of the Statutes of the Commission. Before taking any decision involving expenditure in connection

with the establishment of such subsidiary bodies, the Commission shall have before it a report from the Director-General of FAO and/or WHO, as appropriate, on the administrative and financial implications thereof.

10. The Members who shall be responsible for appointing Chairpersons of subsidiary bodies established under Rule IX.1(b)(i) shall be designated at each session by the Commission, except where otherwise provided in these Rules, and shall be eligible for re-designation. All other officers of subsidiary bodies shall be elected by the body concerned and shall be eligible for re-election.

11. The Rules of Procedure of the Commission shall apply *mutatis mutandis* to its subsidiary bodies.

RULE X ELABORATION OF STANDARDS

1. Subject to the provisions of these Rules of Procedure, the Commission may establish the procedures for the elaboration of world wide standards and of standards for a given region or group of countries, and, when necessary, amend such procedures.

RULE XI BUDGET AND EXPENSES

1. The Directors General of FAO and WHO shall prepare for consideration by the Commission at its regular sessions an estimate of expenditure based on the proposed programme of work of the Commission and its subsidiary bodies, together with information concerning expenditures for the previous financial period. This estimate, with such modifications as may be considered appropriate by the Directors General in the light of recommendations made by the Commission, shall subsequently be incorporated in the Regular Budgets of the two Organizations for approval by the appropriate governing bodies.

2. The estimate of expenditure shall make provisions for the operating expenses of the Commission and the subsidiary bodies of the Commission established under Rule IX.1(a) and IX.1(b)(ii) and for the expenses relating to staff assigned to the Programme and other expenditures incurred in connection with the servicing of the latter.

3. The operating costs of subsidiary bodies established under Rule IX.1(b)(i) (Codex Committees) shall be borne by each Member accepting the Chair of such a body. The estimate of expenditure may include a provision for such costs involved in preparatory work as may be recognized as operating expenses of the Commission in accordance with the provisions of Article 10 of the Statutes of the Commission.

4. Expenses incurred in connection with attendance at sessions of the Commission and its subsidiary bodies and travel of delegations of the Members of the Commission and of observers referred to in Rule VII, shall be borne by

the governments or organizations concerned. Should experts be invited by the Directors-General of FAO or WHO to attend sessions of the Commission and its subsidiary bodies in their individual capacity, their expenses shall be borne out of the regular budgetary funds available for the work of the Commission.

RULE XII LANGUAGES

1. The languages of the Commission and of its subsidiary bodies set up under Rule IX.1(a) shall be not less than three of the working languages, as shall be determined by the Commission, which are working languages both of FAO and of the Health Assembly of WHO.

2. Notwithstanding the provisions of paragraph 1 above, other languages which are working languages either of FAO or of the Health Assembly of WHO may be added by the Commission if

(a) the Commission has before it a report from the Directors General of FAO and WHO on the policy, financial and administrative implications of the addition of such languages; and

(b) the addition of such languages has the approval of the Directors General of FAO and WHO.

3. Where a representative wishes to use a language other than a language of the Commission he shall himself provide the necessary interpretation and/or translation into one of the languages of the Commission.

4. Without prejudice to the provisions of paragraph 3 of this Rule, the languages of subsidiary bodies set up under Rule IX.1(b) shall include at least two of the languages of the Commission.

RULE XIII AMENDMENTS AND SUSPENSION OF RULES

1. Amendments of or additions to these Rules may be adopted by a two thirds majority of the votes cast, provided that 24 hours' notice of the proposal for the amendment or addition has been given. Amendments of or additions to these Rules shall come into force upon approval by the Directors General of FAO and WHO, subject to such confirmation as may be prescribed by the procedures of the two Organizations.

2. The Rules of the Commission, other than Rule I, Rule II.1, 2, 3 and 6, Rule III, Rule IV.2 and 6, Rule V.1, 4 and 6, Rule VI.1, 2 and 3, Rule VII, Rule VIII.3 and 4, Rule IX.5, 7 and 9, Rule XI, Rule XIII and Rule XIV, may be suspended by the Commission by a two thirds majority of the votes cast, provided that 24 hours' notice of the proposal for suspension has been given. Such notice may be waived if no representative of the Members of the Commission objects.

RULE XIV   ENTRY INTO FORCE

1.   In accordance with Article 8 of the Statutes of the Commission, these Rules of Procedure shall come into force upon approval by the Directors General of FAO and WHO, subject to such confirmation as may be prescribed by the procedures of the two Organizations. Pending the coming into force of these Rules, they shall apply provisionally.

PROCEDURES FOR THE ELABORATION OF CODEX STANDARDS AND RELATED TEXTS

Note: Throughout this text the word "Standard" is meant to include any of the recommendations of the Commission intended to be submitted to Governments for acceptance. Except for provisions relating to acceptance, the Procedures apply *mutatis mutandis* to codes of practice and other texts of an advisory nature.

INTRODUCTION

1. The full procedure for the elaboration of Codex standards is as follows. The Commission decides, taking into account the "Criteria for the Establishment of Work Priorities and for the Establishment of Subsidiary Bodies", that a standard should be elaborated and also which subsidiary body or other body should undertake the work. Decisions to elaborate standards may also be taken by subsidiary bodies of the Commission in accordance with the above-mentioned criteria subject to subsequent approval by the Commission or its Executive Committee at the earliest possible opportunity. The Secretariat arranges for the preparation of a "proposed draft standard" which is circulated to governments for comments and is then considered in the light of these by the subsidiary body concerned which may present the text to the Commission as a "draft standard". If the Commission adopts the "draft standard" it is sent to governments for further comments and in the light of these and after further consideration by the subsidiary body concerned, the Commission reconsiders the draft and may adopt it as a "Codex standard". The procedure is described in Part 1 of this document.

2. The Commission or the Executive Committee, or any subsidiary body, subject to the confirmation of the Commission or the Executive Committee may decide that the urgency of elaborating a Codex Standard is such that an accelerated elaboration procedure should be followed. While taking this decision, all appropriate matters shall be taken into consideration, including the likelihood of new scientific information becoming available in the immediate future. The accelerated elaboration procedure is described in Part 2 of this document.

3. The Commission or the subsidiary body or other body concerned may decide that the draft be returned for further work at any appropriate previous Step in the Procedure. The Commission may also decide that the draft be held at Step 8.

4. The Commission may authorize, on the basis of two-thirds majority of votes cast, the omission of Steps 6 and 7, where such an omission is recommended by the Codex Committee entrusted with the elaboration of the draft. Recommendations to omit steps shall be notified to Members and interested international organizations as soon as possible after the session of the Codex Committee concerned. When formulating recommendations to omit Steps 6 and 7, Codex Committees shall take all appropriate matters into consideration, including the need for urgency, and the likelihood of new scientific information becoming available in the immediate future.

5. The Commission may at any stage in the elaboration of a standard entrust any of the remaining Steps to a Codex Committee or other body different from that to which it was previously entrusted.

6. It will be for the Commission itself to keep under review the revision of "Codex standards". The procedure for revision should, *mutatis mutandis*, be that laid down for the elaboration of Codex standards, except that the Commission may decide to omit any other step or steps of that Procedure where, in its opinion, an amendment proposed by a Codex Committee is either of an editorial nature or of a substantive nature but consequential to provisions in similar standards adopted by the Commission at Step 8.

7. Codex standards are published and sent to governments for acceptance. They are also sent to international organizations to which competence in the matter has been transferred by their Member States. See Part 3 of this document. Details of Government acceptances are published periodically by the Commission's Secretariat.

PART 1: UNIFORM PROCEDURE FOR THE ELABORATION OF CODEX STANDARDS AND RELATED TEXTS

STEPS 1, 2 AND 3

(1) The Commission decides, taking into account the "Criteria for the Establishment of Work Priorities and for the Establishment of Subsidiary Bodies", to elaborate a Worldwide Codex Standard and also decides which subsidiary body or other body should undertake the work. A decision to elaborate a Worldwide Codex Standard may also be taken by subsidiary bodies of the Commission in accordance with the above mentioned criteria, subject to subsequent approval by the Commission or its Executive Committee at the earliest possible opportunity. In the case of Codex Regional Standards, the Commission shall base its decision on the proposal of the majority of Members belonging to a given region or group of countries submitted at a session of the Codex Alimentarius Commission.

(2) The Secretariat arranges for the preparation of a proposed draft standard. In the case of Maximum Limits for Residues of Pesticides or Veterinary Drugs, the Secretariat distributes the recommendations for maximum limits, when available from the Joint Meetings of the FAO Panel of Experts on Pesticide Residues in Food and the Environment and the WHO Panel of Experts on Pesticide Residues (JMPR), or the Joint FAO/WHO Expert Committee on Food Additives (JECFA). In the cases of milk and milk products or individual standards for cheeses, the Secretariat distributes the recommendations of the International Dairy Federation (IDF).

(3) The proposed draft standard is sent to Members of the Commission and interested international organizations for comment on all aspects including possible implications of the proposed draft standard for their economic interests.

STEP 4

The comments received are sent by the Secretariat to the subsidiary body or other body concerned which has the power to consider such comments and to amend the proposed draft standard.

STEP 5

The proposed draft standard is submitted through the Secretariat to the Commission or to the Executive Committee with a view to its adoption as a draft standard.[1] In taking any decision at this step, the Commission or the Executive Committee will give due consideration to any comments that may be submitted by any of its Members regarding the implications which the proposed draft standard or any provisions thereof may have for their economic interests. In the case of Regional Standards, all Members of the Commission may present their comments, take part in the debate and propose amendments, but only the majority of the Members of the region or group of countries concerned attending the session can decide to amend or adopt the draft. In taking any decisions at this step, the Members of the region or group of countries concerned will give due consideration to any comments that may be submitted by any of the Members of the Commission regarding the implications which the proposed draft standard or any provisions thereof may have for their economic interests.

STEP 6

The draft standard is sent by the Secretariat to all Members and interested international organizations for comment on all aspects, including possible implications of the draft standard for their economic interests.

STEP 7

The comments received are sent by the Secretariat to the subsidiary body or other body concerned, which has the power to consider such comments and amend the draft standard.

STEP 8

The draft standard is submitted through the Secretariat to the Commission together with any written proposals received from Members and interested international organizations for amendments at Step 8 with a view to its adoption as a Codex standard. In the case of Regional standards, all Members and interested international organizations may present their comments, take part in the debate and propose amendments but only the majority of Members of the region or group of countries concerned attending the session can decide to amend and adopt the draft.

---

[1] Without prejudice to any decision that may be taken by the Commission at Step 5, the proposed draft standard may be sent by the Secretariat for government comments prior to its consideration at Step 5, when, in the opinion of the subsidiary body or other body concerned, the time between the relevant session of the Commission and the subsequent session of the subsidiary body or other body concerned requires such action in order to advance the work.

PART 2: UNIFORM ACCELERATED PROCEDURE FOR THE ELABORATION OF CODEX STANDARDS AND RELATED TEXTS

STEPS 1, 2 AND 3

(1) The Commission or the Executive Committee between Commission sessions, on the basis of a two-thirds majority of votes cast, taking into account the "Criteria for the Establishment of Work Priorities and for the Establishment of Subsidiary Bodies", shall identify those standards which shall be the subject of an accelerated elaboration process.[1] The identification of such standards may also be made by subsidiary bodies of the Commission, on the basis of a two-thirds majority of votes cast, subject to confirmation at the earliest opportunity by the Commission or its Executive Committee by a two-thirds majority of votes cast.

(2) The Secretariat arranges for the preparation of a proposed draft standard. In the case of Maximum Limits for Residues of Pesticides or Veterinary Drugs, the Secretariat distributes the recommendations for maximum limits, when available from the Joint Meetings of the FAO Panel of Experts on Pesticide Residues in Food and the Environment and the WHO Panel of Experts on Pesticide Residues (JMPR), or the Joint FAO/WHO Expert Committee on Food Additives (JECFA). In the cases of milk and milk products or individual standards for cheeses, the Secretariat distributes the recommendations of the International Dairy Federation (IDF).

(3) The proposed draft standard is sent to Members of the Commission and interested international organizations for comment on all aspects including possible implications of the proposed draft standard for their economic interests. When standards are subject to an accelerated procedure, this fact shall be notified to the Members of the Commission and the interested international organizations.

STEP 4

The comments received are sent by the Secretariat to the subsidiary body or other body concerned which has the power to consider such comments and to amend the proposed draft standard.

---

[1] Relevant considerations could include, but need not be limited to, matters concerning new scientific information; new technology(ies); urgent problems related to trade or public health; or the revision or up-dating of existing standards.

STEP 5

In the case of standards identified as being subject to an accelerated elaboration procedure, the draft standard is submitted through the Secretariat to the Commission together with any written proposals received from Members and interested international organizations for amendments with a view to its adoption as a Codex standard. In taking any decision at this step, the Commission will give due consideration to any comments that may be submitted by any of its Members regarding the implications which the proposed draft standard or any provisions thereof may have for their economic interests.

PART 3: SUBSEQUENT PROCEDURE CONCERNING PUBLICATION AND ACCEPTANCE OF CODEX STANDARDS

The Codex standard is published and issued to all Member States and Associate Members of FAO and/or WHO and to the international organizations concerned. Members of the Commission and international organizations to which competence in the matter has been transferred by their Member States notify the Secretariat of their acceptance of the Codex standard in accordance with the acceptance procedure laid down in paragraph 4, paragraph 5 or in paragraph 6 of the General Principles of the Codex Alimentarius, whichever is appropriate. Member States and Associate Members of FAO and/or WHO that are not Members of the Commission are invited to notify the Secretariat if they wish to accept the Codex standard.

The Secretariat publishes periodically details of notifications received from governments and from international organizations to which competence in the matter has been transferred by their Member States with respect to the acceptance or otherwise of Codex standards and in addition to this information an appendix for each Codex standard (a) listing the countries in which products conforming with such standard may be freely distributed, and (b) where applicable, stating in detail all specified deviations which may have been declared in respect to the acceptance.

The above mentioned publications will constitute the *Codex Alimentarius*.

The Secretariat examines deviations notified by governments and reports periodically to the Codex Alimentarius Commission concerning possible amendments to standards which might be considered by the Commission in accordance with the Procedure for the Revision and Amendment of Recommended Codex Standards.

SUBSEQUENT PROCEDURE CONCERNING PUBLICATION, ACCEPTANCE AND POSSIBLE EXTENSION OF TERRITORIAL APPLICATION OF THE STANDARD

The Codex Regional Standard is published and issued to all Member States and Associate Members of FAO and/or WHO and to the international organizations concerned. Members of the region or group of countries concerned notify the Secretariat of their acceptance of the Codex Regional Standard in accordance with the acceptance procedure laid down in paragraph 4, paragraph 5 or in paragraph 6 of the General Principles of the Codex Alimentarius, whichever is appropriate. Other Members of the Commission may likewise notify the Secretariat of their acceptance of the standard or of any other measures they propose to adopt with respect thereto, and also submit any observations as to its application. Member States and Associate Members of FAO and/or WHO that are not Members of the Commission are invited to notify the Secretariat if they wish to accept the standard.

It is open to the Commission to consider at any time the possible extension of the territorial application of a Codex Regional Standard or its conversion into a Worldwide Codex Standard in the light of all acceptances received.

GUIDE TO THE CONSIDERATION OF STANDARDS AT STEP 8 OF THE PROCEDURE FOR THE ELABORATION OF CODEX STANDARDS INCLUDING CONSIDERATION OF ANY STATEMENTS RELATING TO ECONOMIC IMPACT

1. In order:

    (a) to ensure that the work of the Codex committee concerned is not made less valuable by the passage of an insufficiently considered amendment in the Commission;

    (b) at the same time to provide scope for significant amendments to be raised and considered in the Commission;

    (c) to prevent, as far as practicable, lengthy discussion in the Commission on points that have been thoroughly argued in the Codex committee concerned;

    (d) to ensure, as far as practicable, that delegations are given sufficient warning of amendments so that they may brief themselves adequately,

    amendments to Codex standards at Step 8 should, as far as practicable, be submitted in writing, although amendments proposed in the Commission would not be excluded entirely, and the following procedure should be employed:

2. When Codex standards are distributed to Member Countries prior to their consideration by the Commission at Step 8, the Secretariat will indicate the date by which proposed amendments must be received; this date will be fixed so as to allow sufficient time for such amendments to be in the hands of governments not less than one month before the session of the Commission.

3. Governments should submit amendments in writing by the date indicated and should state that they had been previously submitted to the appropriate Codex committee with details of the submission of the amendment or should give the reason why the amendment had not been proposed earlier, as the case may be.

4. When amendments are proposed during a session of the Commission, without prior notice, to a standard which is at Step 8, the Chairperson of the Commission, after consultation with the chairperson of the appropriate committee, or, if the chairperson is not present, with the delegate of the chairing country, or, in the case of subsidiary bodies which do not have a chairing country, with other appropriate persons, shall rule whether such amendments are substantive.

5. If an amendment ruled as substantive is agreed to by the Commission, it shall be referred to the appropriate Codex committee for its comments and, until such comments have been received and considered by the Commission, the standard shall not be advanced beyond Step 8 of the Procedure.

6. It will be open to any Member of the Commission to draw to the attention of the Commission any matter concerning the possible implications of a draft standard for its economic interests, including any such matter which has not, in that Member's opinion, been satisfactorily resolved at an earlier step in the Procedure for the Elaboration of Codex Standards. All the information pertaining to the matter, including the outcome of any previous consideration by the Commission or a subsidiary body thereof should be presented in writing to the Commission, together with any draft amendments to the standard which would in the opinion of the country concerned, take into account the economic implications. In considering statements concerning economic implications the Commission should have due regard to the purposes of the Codex Alimentarius concerning the protection of the health of consumers and the ensuring of fair practices in the food trade, as set forth in the General Principles of the Codex Alimentarius, as well as the economic interests of the Member concerned. It will be open to the Commission to take any appropriate action including referring the matter to the appropriate Codex committee for its comments.

GUIDE TO THE PROCEDURE FOR THE REVISION AND AMENDMENT OF CODEX STANDARDS

1. Proposals for the amendment or revision of Codex standards should be submitted to the Commission's Secretariat in good time (not less than three months) before the session of the Commission at which they are to be considered. The proposer of an amendment should indicate the reasons for the proposed amendment and should also state whether the proposed amendment had been previously submitted to and considered by the Codex committee concerned and/or the Commission. If the proposed amendment has already been considered by the Codex committee and/or Commission, the outcome of the consideration of the proposed amendment should be stated.

2. Taking into account such information regarding the proposed amendment as may be supplied in accordance with paragraph 1 above, the Commission will decide whether the amendment or revision of a standard is necessary. If the Commission decides in the affirmative, and the proposer of the amendment is other than a Codex committee, the proposed amendment will be referred for consideration to the appropriate Codex committee, if such committee is still in existence. If such committee is not in existence, the Commission will determine how best to deal with the proposed amendment. If the proposer of the amendment is a Codex committee, it would be open to the Commission to decide that the proposed amendment be circulated to governments for comments prior to further consideration by the sponsoring Codex Committee. In the case of an amendment proposed by a Codex Committee, it will also be open to the Commission to adopt the amendment at Step 5 or Step 8 as appropriate, where in its opinion the amendment is either of an editorial nature or of a substantive nature but consequential to provisions in similar standards adopted by it at Step 8.

3. The procedure for amending or revising a Codex standard would be as laid down in paragraphs 3 and 4 of the Introduction to the Procedure for the Elaboration of Codex Standards (see page 18 above).

4. When the Commission has decided to amend or revise a standard, the unrevised standard will remain the applicable Codex standard until the revised standard has been adopted by the Commission.

ARRANGEMENTS FOR THE AMENDMENT OF CODEX STANDARDS ELABORATED BY CODEX COMMITTEES WHICH HAVE ADJOURNED *SINE DIE*

I. The need to consider amending or revising adopted Codex standards arises from time to time for a variety of reasons amongst which can be:

   A. changes in the evaluation of food additives, pesticides and contaminants;

   B. finalization of methods of analysis;

   C. editorial amendments of guidelines or other texts adopted by the Commission and related to all or a group of Codex standards e.g. "Guidelines on Date Marking", "Guidelines on Labelling of Non retail Containers", "Carry-over Principle";

   D. consequential amendments to earlier Codex standards arising from Commission decisions on currently adopted standards of the same type of products;

   E. consequential and other amendments arising from either revised or newly elaborated Codex standards and other texts of general applicability which have been referenced in other Codex standards (Revision of General Principles of Food Hygiene, Codex Standard for the Labelling of Prepackaged Foods);

   F. technological developments or economic considerations e.g. provisions concerning styles, packaging media or other factors related to composition and essential quality criteria and consequential changes in labelling provisions;

   G. modifications of standards being proposed following an examination of government notifications of acceptances and specified deviations by the Secretariat as required in accordance with the Procedure for the Elaboration of Codex standards i.e. "Subsequent Procedure concerning Publication and Acceptance of Codex Standards", page 23.

II. The Guide to the Procedure for the Revision and Amendment of Codex Standards (see page 26) covers sufficiently amendments to Codex standards which have been elaborated by still active Codex Committees and those mentioned under paragraph 1 (g) above. In the case of amendments proposed to Codex standards elaborated by Codex Committees which have adjourned *sine die*, the procedure places an obligation on the Commission to "determine how best to deal with the proposed amendment". In order to facilitate consideration of such amendments, in particular, those of the type mentioned in para. 1 (a), (b), (c), (d), (e) and (f), the Commission has established more detailed guidance within the existing procedure for the amendment and revision of Codex standards.

III. In the case where Codex committees have adjourned *sine die*:

    A. the Secretariat keep under review all Codex standards originating from Codex Committees adjourned *sine die* and to determine the need for any amendments arising from decisions of the Commission, in particular amendments of the type mentioned in para. 1(a), (b), (c), (d) and those of (e) if of an editorial nature. If a need to amend the standard appears appropriate then the Secretariat should prepare a text for adoption in the Commission;

    B. amendments of the type in para (f) and those of (e) of a substantive nature, the Secretariat in cooperation with the national secretariat of the adjourned Committee and, if possible, the Chairperson of that Committee, should agree on the need for such an amendment and prepare a working paper containing the wording of a proposed amendment and the reasons for proposing such amendment, and request comments from Member Governments: (a) on the need to proceed with such an amendment and (b) on the proposed amendment itself. If the majority of the replies received from Member Governments is affirmative on both the need to amend the standard and the suitability of the proposed wording for the amendment or an alternative proposed wording, the proposal should be submitted to the Commission with a request to approve the amendment of the standard concerned. In cases where replies do not appear to offer an uncontroversial solution then the Commission should be informed accordingly and it would be for the Commission to determine how best to proceed.

GENERAL PRINCIPLES OF THE CODEX ALIMENTARIUS

PURPOSE OF THE CODEX ALIMENTARIUS

1. The Codex Alimentarius is a collection of internationally adopted food standards presented in a uniform manner. These food standards aim at protecting consumers' health and ensuring fair practices in the food trade. The Codex Alimentarius also includes provisions of an advisory nature in the form of codes of practice, guidelines and other recommended measures intended to assist in achieving the purposes of the Codex Alimentarius. The publication of the Codex Alimentarius is intended to guide and promote the elaboration and establishment of definitions and requirements for foods to assist in their harmonization and in doing so to facilitate international trade.

SCOPE OF THE CODEX ALIMENTARIUS

2. The Codex Alimentarius includes standards for all the principle foods, whether processed, semi-processed or raw, for distribution to the consumer. Materials for further processing into foods should be included to the extent necessary to achieve the purposes of the Codex Alimentarius as defined. The Codex Alimentarius includes provisions in respect of food hygiene, food additives, pesticide residues, contaminants, labelling and presentation, methods of analysis and sampling. It also includes provisions of an advisory nature in the form of codes of practice, guidelines and other recommended measures.

NATURE OF CODEX STANDARDS

3. Codex standards contain requirements for food aimed at ensuring for the consumer a sound, wholesome food product free from adulteration, correctly labelled and presented. A Codex standard for any food or foods should be drawn up in accordance with the Format for Codex Commodity Standards and contain, as appropriate, the criteria listed therein.

ACCEPTANCE OF CODEX COMMODITY STANDARDS

4.A. A Codex standard may be accepted by a country in accordance with its established legal and administrative procedures in respect of distribution of the product concerned, whether imported or home produced, within its territorial jurisdiction in the following ways:

    (i) Full acceptance

    (a) Full acceptance means that the country concerned will ensure that a product to which the standard applies will be permitted to be distributed freely, in accordance with (c) below, within its territorial jurisdiction under the name and description laid down in the standard, provided that it complies with all the relevant requirements of the standard.

(b) The country will also ensure that products not complying with the standard will not be permitted to be distributed under the name and description laid down in the standard.

(c) The distribution of any sound products conforming with the standard will not be hindered by any legal or administrative provisions in the country concerned relating to the health of the consumer or to other food standard matters except for considerations of human, plant or animal health which are not specifically dealt with in the standard.

(ii) Acceptance with specified deviations

Acceptance with specified deviations means that the country concerned gives acceptance, as defined in paragraph 4.A(i), to the standard with the exception of such deviations as are specified in detail in its declaration of acceptance; it being understood that a product complying with the standard as qualified by these deviations will be permitted to be distributed freely within the territorial jurisdiction of the country concerned. The country concerned will further include in its declaration of acceptance a statement of the reasons for these deviations, and also indicate:

(a) whether products fully conforming to the standard may be distributed freely within its territorial jurisdiction in accordance with paragraph 4.A(i);

(b) whether it expects to be able to give full acceptance to the standard and, if so, when.

(iii) Free distribution

A declaration of free distribution means that the country concerned undertakes that products conforming with a Codex commodity standard may be distributed freely within its territorial jurisdiction insofar as matters covered by the Codex commodity standard are concerned.

B. A country which considers that it cannot accept the standard in any of the ways mentioned above should indicate:

(i) whether products conforming to the standard may be distributed freely within its territorial jurisdiction;

(ii) in what ways its present or proposed requirements differ from the standard, and, if possible the reasons for these differences.

C. (i) A country which accepts a Codex standard according to one of the provisions of 4.A is responsible for the uniform and impartial application of the provisions of the standard as accepted, in respect of all home-produced and imported products distributed within its territorial jurisdiction. In addition, the country should be prepared to offer advice and guidance to exporters and processors of products for export to promote understanding of and compliance with the requirements of importing countries which have accepted a Codex standard according to one of the provisions of 4.A.

(ii) Where, in an importing country, a product claimed to be in compliance with a Codex standard is found not to be in compliance with that standard, whether in respect of the label accompanying the product or otherwise, the importing country should inform the competent authorities in the exporting country of all the relevant facts and in particular the details of the origin of the product in question (name and address of the exporter), if it is thought that a person in the exporting country is responsible for such non-compliance.

ACCEPTANCE OF CODEX GENERAL STANDARDS

5.A. A Codex general standard may be accepted by a country in accordance with its established legal and administrative procedures in respect of the distribution of products to which the general standard applies, whether imported or home-produced, within its territorial jurisdiction in the following ways:

(i) Full acceptance

Full acceptance of a general standard means that the country concerned will ensure, within its territorial jurisdiction, that a product to which the general standard applies will comply with all the relevant requirements of the general standard except as otherwise provided in a Codex commodity standard. It also means that the distribution of any sound products conforming with the standard will not be hindered by any legal or administrative provisions in the country concerned, which relate to the health of the consumer or to other food standard matters and which are covered by the requirements of the general standard.

(ii) Acceptance with specified deviations

Acceptance with specified deviations means that the country concerned gives acceptance, as defined in paragraph 5.A(i), to the general standard with the exception of such deviations as are specified in detail in its declaration of acceptance. The country concerned will further include in its declaration of acceptance a statement of the reasons for these deviations, and also indicate whether it expects to be able to give full acceptance to the general standard and, if so, when.

(iii) Free distribution

A declaration of free distribution means that the country concerned undertakes that products conforming with the relevant requirements of a Codex general standard may be distributed freely within its territorial jurisdiction insofar as matters covered by the Codex general standard are concerned.

B. A country which considers that it cannot accept the general standard in any of the ways mentioned above should indicate in what ways its present or proposed requirements differ from the general standard, and if possible, the reasons for these differences.

C. (i) A country which accepts a general standard according to one of the provisions of paragraph 5.A is responsible for the uniform and impartial application of the provisions of the standard as accepted, in respect of all home produced and imported products distributed within its territorial jurisdiction. In addition, the country should be prepared to offer advice and guidance to exporters and processors of products for export to promote understanding of and compliance with the requirements of importing countries which have accepted a general standard according to one of the provisions of paragraph 5.A.

(ii) Where, in an importing country, a product claimed to be in compliance with a general standard is found not to be in compliance with that standard, whether in respect of the label accompanying the product or otherwise, the importing country should inform the competent authorities in the exporting country of all the relevant facts and in particular the details of the origin of the product in question (name and address of the exporter), if it is thought that a person in the exporting country is responsible for such non-compliance.

ACCEPTANCE OF CODEX MAXIMUM LIMITS FOR RESIDUES OF PESTICIDES AND VETERINARY DRUGS IN FOOD

6.A .A Codex maximum limit for residues of pesticides or veterinary drugs in food may be accepted by a country in accordance with its established legal and administrative procedures in respect of the distribution within its territorial jurisdiction of (a) home-produced and imported food or (b) imported food only, to which the Codex maximum limit applies in the ways set forth below. In addition, where a Codex maximum limit applies to a group of foods not individually named, a country accepting such Codex maximum limit in respect of other than the group of foods, shall specify the foods in respect of which the Codex maximum limit is accepted.

(i) Full acceptance

Full acceptance of a Codex maximum limit for residues of pesticides or veterinary drugs in food means that the country concerned will ensure, within its territorial jurisdiction, that a food, whether home-produced or imported, to which the Codex maximum limit applies, will comply with that limit. It also means that the distribution of a food conforming with the Codex maximum limit will not be hindered by any legal or administrative provisions in the country concerned which relate to matters covered by the Codex maximum limit.

(ii) Free distribution

A declaration of free distribution means that the country concerned undertakes that products conforming with the Codex maximum limit for residues of pesticides or veterinary drugs in food may be distributed freely within its territorial jurisdiction insofar as matters covered by the Codex maximum limit are concerned.

B. A country which considers that it cannot accept the Codex maximum limit for residues of pesticides or veterinary drugs in foods in any of the ways mentioned above should indicate in what ways its present or proposed requirements differ from the Codex maximum limit and, if possible, the reasons for these differences.

C. A country which accepts a Codex maximum limit for residues of pesticides or veterinary drugs in food according to one of the provisions of paragraph 6.A should be prepared to offer advice and guidance to exporters and processors of food for export to promote understanding of and compliance with the requirements of importing countries which have accepted a Codex maximum limit according to one of the provisions of paragraph 6.A.

D. Where, in an importing country, a food claimed to be in compliance with a Codex maximum limit is found not to be in compliance with the Codex maximum limit, the importing country should inform the competent authorities in the exporting country of all the relevant facts and, in particular, the details of the origin of the food in question (name and address of the exporter), if it is thought that a person in the exporting country is responsible for such non-compliance.

WITHDRAWAL OR AMENDMENT OF ACCEPTANCE

7. The withdrawal or amendment of acceptance of a Codex standard or a Codex maximum limit for residues of pesticides or veterinary drugs in food by a country shall be notified in writing to the Codex Alimentarius Commission's Secretariat who will inform all Member States and Associate Members of FAO and WHO of the notification and its date of receipt. The country concerned should provide the information required under paragraphs 4.A(iii), 5.A(iii), 4.B, 5.B or 6.B above, whichever is appropriate. It should also give as long a notice of the withdrawal or amendment as is practicable.

REVISION OF CODEX STANDARDS

8. The Codex Alimentarius Commission and its subsidiary bodies are committed to revision as necessary of Codex standards and related texts to ensure that they are consistent with and reflect current scientific knowledge and other relevant information. When required, a standard or related text shall be revised or removed using the same procedures as followed for the elaboration of a new standard. Each member of the Codex Alimentarius Commission is responsible for identifying, and presenting to the appropriate committee, any new scientific and other relevant information which may warrant revision of any existing Codex standards or related texts.

GUIDELINES FOR THE ACCEPTANCE PROCEDURE FOR CODEX STANDARDS

THE IMPORTANCE OF A RESPONSE TO EVERY NOTIFICATION

1. The Codex Alimentarius is the record of Codex Standards and of acceptances or other notifications by Member Countries or international organizations to which competence in the matter has been transferred by their Member States. It is revised regularly to take account of the issue of new or amended standards and the receipt of notifications. It is important that governments respond to every issue of new or amended standards. Governments should aim at giving formal acceptance to the standards. If acceptance or free circulation cannot be given unconditionally, the deviations or conditions, and the reasons, can be included in the response. Early and regular responses will ensure that the Codex Alimentarius can be kept up to date so as to serve as an indispensable reference for governments and international traders.

2. Governments should ensure that the information in the Codex Alimentarius reflects the up to date position. When changing national laws or practices the need for a notification to the Codex Secretariat should always be kept in mind.

3. The Codex procedure for elaboration of standards enables governments to participate at all stages.. Governments should be able to make an early response to the issue of a Codex standard and should do their utmost to be ready to do so.

THE CODEX ALIMENTARIUS: NOT A SUBSTITUTE FOR, OR ALTERNATIVE TO, REFERRING TO NATIONAL LEGISLATION

4. Every country's laws and administrative procedures contain provisions which it is essential to understand and comply with. It is usually the practice to take steps to obtain copies of relevant legislation and/or to obtain professional advice about compliance. The Codex Alimentarius is a comparative record of the substantive similarities and differences between Codex Standards and corresponding national legislation. The Codex Standard will not normally deal with general matters of human, plant or animal health or with trade marks. The language which is required on labels will be a matter for national legislation and so will import licences and other administrative procedures.

5. The responses by governments should show clearly which provisions of the Codex Standard are identical to, similar to or different from, the related

national requirements. General statements that national laws must be complied with should be avoided or accompanied by details of national provisions which require attention. Judgement will sometimes be required where the national law is in a different form or where it has different provisions.

OBLIGATIONS UNDER THE ACCEPTANCE PROCEDURE

6. The obligations which a country undertakes under the acceptance procedure are included in paragraph 4 of the General Principles. Paragraph 4A(i)(a) provides for free distribution of conforming products, 4A(i)(b) with the need to ensure that products which do not conform may not be distributed "under the name and description laid down". Paragraph 4A(i)(c) is a general requirement not to hinder the distribution of sound products, except for matters relating to human, plant or animal health, not specifically dealt with in the standard. Similar provisions are included in Acceptance with Specified Deviations.

7. The essential difference between acceptances and notifications of free distribution is that a country which accepts, undertakes to enforce the Codex standard and to accept all the obligations set out in the General Principles subject to any specified deviations.

8. The Codex Committee on General Principles (CCGP) and the Commission (CAC) have reviewed the acceptance procedure and notifications by governments on a number of occasions. While recognizing that difficulties can arise from time to time in reconciling the obligations of the acceptance procedure with the laws and administrative procedures of a Member Country, the CCGP and the CAC have determined that the obligations are essential to the work and status of the CAC and that they should not be weakened in any way. The purpose of these guidelines therefore is to assist governments when they are considering how, in the light of the objectives of the acceptance procedure, to respond to Codex Standards.

THE RETURN OF THE RESPONSE

9. The principal decision which is required is whether to notify an acceptance according to one of the methods prescribed, or non acceptance as provided for in 4B. Free distribution (4A(iii)) does not carry with it the obligation to prevent non conforming products from being circulated, and it may be useful in cases where there is no corresponding national standard and no intention to introduce one.

THE NEED FOR AN INFORMED, RESPONSIBLE JUDGEMENT WHEN COMPARING THE CODEX STANDARD WITH NATIONAL LAWS

10. There will be some occasions when the detail in the Codex Standard is identical with national laws. Difficulties will arise however when national laws are in a different form, contain different figures or no figures at all, or in cases where there may be no standard in the country which corresponds in substance to the Codex Standard. The authority responsible for notifying the response to the CAC is urged to do its best to overcome any such difficulties by the exercise of its best endeavours and to respond, after such consultations as may be appropriate with the national organizations. The grounds on which the judgement has been based can be made clear in the notification. It may well be that they will not be such as to justify an acceptance, because of the obligations to stop the distribution of non conforming products, but a statement of free circulation should be possible on the basis of the facts and practices of each case. If there was a court decision or change in the law or practice subsequently, an amending response should be made.

PRESUMPTIVE STANDARDS

11. A presumptive standard is one which is assumed to be the standard in the absence of any other. (A presumption in law is the assumption of the truth of anything until the contrary is proved.) Some countries have said that a Codex MRL is the presumptive limit for a pesticide residue. Countries may be able and willing to regard a Codex Standard as the presumptive standard in cases where there is no corresponding standard, code of practice or other accepted expression of the "nature, substance or quality" of the food. A country need not apply the presumption to all the provisions of the standard if the details of its additives, contaminants, hygiene or labelling rules are different from those in the standard. In such a case the provisions in the Codex Standard defining the description, essential composition and quality factors relating to the specified name and description could still be the presumptive standard for those matters.

12. The justification for regarding the Codex Standard as a presumptive standard is the fact that it is the minimum standard for a food elaborated in the CAC "so as to ensure a sound, wholesome product free from adulteration, correctly labelled and presented". (General Principles, Paragraph 3.) The word minimum does not have any pejorative connotations: it simply means the level of quality and soundness of a product judged by consensus to be appropriate for trade internationally and nationally.

13. Whether a presumptive standard would merit an acceptance would depend on whether the country concerned could say that non conforming products could not be distributed under the same name and description laid

down in the standard. However it would enable a declaration of free circulation to be made and countries are asked to give the idea serious consideration.

FORMAT AND CONTENT OF CODEX STANDARDS

SCOPE

14. This section, together with the name of the standard and the name and description laid down in the labelling section, should be examined in order to assess whether the obligations of the acceptance procedure can properly be accepted.

DESCRIPTION, ESSENTIAL COMPOSITION AND QUALITY FACTORS

15. These sections will define the minimum standard for the food. They will be the most difficult to address unless by chance the details are virtually identical (i.e. ignoring significant matters of editorial expression or format). However, a country which has taken part in the elaboration of the standard either by attending the meetings or by sending comments under the Step Procedure has, no doubt, consulted national organizations on the extent to which the draft provisions in the standard would be acceptable nationally. This factual information needs to be turned into a formal response when the standard is sent out for acceptance. Countries are asked to do their best to exercise an informal judgement on lines discussed in Paragraph 7 above. Some of the quality criteria e.g. allowances for defects may represent good manufacturing practice or be left to trade contracts. This will have to be taken into account. A free distribution response ought to be possible in most cases.

FOOD ADDITIVES

16. The food additives included in the standard have been assessed and cleared by JECFA. The Commodity Committee and the CCFAC have assessed technological need and safety in use. If national laws are different, all the detailed differences should be reported. It should be borne in mind, however, that the aim of international food standardization work is to harmonize policies and attitudes as much as possible. Therefore every effort should be made to keep deviations to the minimum.

CONTAMINANTS

17. If national limits apply they should be quoted if not the same as those laid down in the Codex Standard. Where general laws about safety, health or nature of the food apply, the limits quoted in the standard could properly be regarded as representing those which are unavoidable in practice and within safety limits.

HYGIENE AND WEIGHTS AND MEASURES

18. If national requirements are different they should be reported.

LABELLING

19. The General Standard for the Labelling of Prepackaged Foods represents the international consensus on information to be included on the labels of all foods.

20. Governments are exhorted to use the General Standard as a basis for their national legislation and to keep differences to an absolute minimum especially those of detail or minutiae. Governments should observe the footnote to the Scope section and should ensure that all compulsory provisions relating to presentation of information which are additional to, and different from, those in the standard should be notified. Any other compulsory provisions in national legislation should also be notified if they are not provided for in the Codex standard. The labelling provisions in Codex standards include sections of the revised General Standard by reference. When accepting a Codex commodity standard, a country which has already accepted and responded to the General Standard can then refer to the terms of that acceptance in any subsequent responses. As much specific information as is relevant and helpful should be given. In particular, this should include the name and description relating to the food, the interpretation of any special requirements relating to the law or custom of the country, any additional details about presentation of the mandatory information and detailed differences if any in the labelling requirements e.g. in relation to class names, declaration of added water, declaration of origin. It will be assumed that the language(s) in which the particulars should be given will be as indicated by national legislation or custom.

METHODS OF ANALYSIS AND SAMPLING

21. The obligations which a country assumes in accepting the following Codex Defining Methods of Analysis included in Codex standards are as follows[1]:

> (a) Codex Defining Methods of Analysis (Type I) are subject to acceptance by governments just as are the provisions which they define and which form part of Codex standards.

---

[1] The Committee on General Principles, when elaborating these Guidelines, noted that the Classification of Methods was under review by the Codex Committee on Methods of Analysis and Sampling and that the application of Part (b) particularly could be unnecessarily restrictive.

"Full acceptance" of a Codex Defining Method means the acceptance that the value provided for in a Codex standard is defined by means of the Codex method. In determining compliance with the value in the Codex standard, governments undertake to use the Codex Defining Method, especially in cases of disputes involving the results of analysis.

"Non acceptance" of Codex Defining Method or acceptance of Codex standards with substantive deviations in the Codex Defining Methods means acceptance of the Codex standard with specified deviation.

(b) The "acceptance" of Codex standards containing Codex Reference Methods of Analysis (Type II) means the recognition that Codex Reference Methods are methods the reliability of which has been demonstrated on the basis of internationally acceptable criteria. They are, therefore, obligatory for use, i.e. subject to acceptance by governments, in disputes involving the results of analysis. "Non acceptance" of the Codex Reference Method or acceptance of Codex standards with substantive deviations in the Codex Reference Methods for use in disputes involving methods of analysis, should be taken to mean acceptance of the Codex standard with specified deviation.

(c) The "acceptance" of Codex standards containing Codex Alternative Approved Methods of Analysis (Type III) means the recognition that Codex Alternative Approved Methods are methods the reliability of which has been demonstrated in terms of internationally acceptable criteria. They are recommended for use in food control, inspection or for regulatory purposes.

"Non acceptance" of a Codex Alternative Approved Method does not constitute a deviation from the Codex standard.

(d) Since the reliability of the Tentative Methods (Type IV) has not yet been endorsed by the Codex Committee on Methods of Analysis and Sampling on the basis of the internationally accepted criteria, it follows that they cannot be regarded as final Codex methods. Type IV methods may, eventually become Type I, II or III methods with the resultant implications regarding the acceptance of Codex methods. Type IV methods are, therefore, not recommended as Codex methods until their reliability has been recognized by the CCMAS. They may be included in draft Codex standards or in Codex standards provided their non approved status is clearly indicated.

SUMMARY

22. Governments are urged to respond to every issue of Codex standards. The inclusion of responses in the Codex Alimentarius will enable the CAC and Member Governments to address the question of closer approximation of

international and national requirements. Governments are urged to take the Codex standard fully into consideration when changing their national laws. The Codex Alimentarius will always be an invaluable reference for governments and for international traders although national legislation must always be consulted and complied with.

DEFINITIONS FOR THE PURPOSES OF THE CODEX ALIMENTARIUS

For the purposes of the Codex Alimentarius:

Food means any substance, whether processed, semi-processed or raw, which is intended for human consumption, and includes drink, chewing gum and any substance which has been used in the manufacture, preparation or treatment of "food" but does not include cosmetics or tobacco or substances used only as drugs.

Food hygiene comprises conditions and measures necessary for the production, processing, storage and distribution of food designed to ensure a safe, sound, wholesome product fit for human consumption.

Food additive means any substance not normally consumed as a food by itself and not normally used as a typical ingredient of the food, whether or not it has nutritive value, the intentional addition of which to food for a technological (including organoleptic) purpose in the manufacture, processing, preparation, treatment, packing, packaging, transport or holding of such food results, or may be reasonably expected to result, (directly or indirectly) in it or its by-products becoming a component of or otherwise affecting the characteristics of such foods. The term does not include "contaminants" or substances added to food for maintaining or improving nutritional qualities.

Contaminant means any substance not intentionally added to food, which is present in such food as a result of the production (including operations carried out in crop husbandry, animal husbandry and veterinary medicine), manufacture, processing, preparation, treatment, packing, packaging, transport or holding of such food or as a result of environmental contamination. The term does not include insect fragments, rodent hairs and other extraneous matter.

Pesticide means any substance intended for preventing, destroying, attracting, repelling, or controlling any pest including unwanted species of plants or animals during the production, storage, transport, distribution and processing of food, agricultural commodities, or animal feeds or which may be administered to animals for the control of ectoparasites. The term includes substances intended for use as a plant growth regulator, defoliant, desiccant, fruit thinning agent, or sprouting inhibitor and substances applied to crops either before or after harvest to protect the commodity from deterioration during storage and transport. The term normally excludes fertilizers, plant and animal nutrients, food additives, and animal drugs.

Pesticide Residue means any specified substance in food, agricultural commodities, or animal feed resulting from the use of a pesticide. The term includes any derivatives of a pesticide, such as conversion products, metabolites, reaction products, and impurities considered to be of toxicological significance.

Good Agricultural Practice in the Use of Pesticides (GAP) includes the nationally authorized safe uses of pesticides under actual conditions necessary for effective and reliable pest control. It encompasses a range of levels of pesticide applications up to the highest authorised use, applied in a manner which leaves a residue which is the smallest amount practicable.

>Authorized safe uses are determined at the national level and include nationally registered or recommended uses, which take into account public and occupational health and environmental safety considerations.

>Actual conditions include any stage in the production, storage, transport, distribution and processing of food commodities and animal feed.

Codex maximum limit for pesticide residues (MRLP) is the maximum concentration of a pesticide residue (expressed as mg/kg), recommended by the Codex Alimentarius Commission to be legally permitted in or on food commodities and animal feeds. MRLs are based on GAP data and foods derived from commodities that comply with the respective MRLs are intended to be toxicologically acceptable.

>Codex MRLs, which are primarily intended to apply in international trade, are derived from estimations made by the JMPR following:

>a)   toxicological assessment of the pesticide and its residue; and

>b)   review of residue data from supervised trials and supervised uses including those reflecting national good agricultural practices. Data from supervised trials conducted at the highest nationally recommended, authorized or registered uses are included in the review. In order to accommodate variations in national pest control requirements, Codex MRLs take into account the higher levels shown to arise in such supervised trials, which are considered to represent effective pest control practices.

>Consideration of the various dietary residue intake estimates and determinations both at the national and international level in comparison with the ADI, should indicate that foods complying with Codex MRLs are safe for human consumption.

Veterinary drug means any substance applied or administered to any food producing animal, such as meat or milk producing animals, poultry, fish or bees, whether used for therapeutic, prophylactic or diagnostic purposes or for modification of physiological functions or behaviour.

Residues of veterinary drugs include the parent compounds and/or their metabolites in any edible portion of the animal product, and include residues of associated impurities of the veterinary drug concerned.

Codex maximum limit for residues of veterinary drugs (MRLVD) is the maximum concentration of residue resulting from the use of a veterinary drug (expressed in mg/kg or µg/kg on a fresh weight basis) that is recommended by the Codex Alimentarius Commission to be legally permitted or recognized as acceptable in or on a food.

> It is based on the type and amount of residue considered to be without any toxicological hazard for human health as expressed by the Acceptable Daily Intake (ADI), or on the basis of a temporary ADI that utilizes an additional safety factor. It also takes into account other relevant public health risks as well as food technological aspects.

> When establishing an MRL, consideration is also given to residues that occur in food of plant origin and/or the environment. Furthermore, the MRL may be reduced to be consistent with good practices in the use of veterinary drugs and to the extent that practical analytical methods are available.

Good Practice in the Use of Veterinary Drugs (GPVD) is the official recommended or authorized usage including withdrawal periods, approved by national authorities, of veterinary drugs under practical conditions.

Processing aid means any substance or material, not including apparatus or utensils, and not consumed as a food ingredient by itself, intentionally used in the processing of raw materials, foods or its ingredients, to fulfil a certain technological purpose during treatment or processing and which may result in the non-intentional but unavoidable presence of residues or derivatives in the final product.

DEFINITIONS OF RISK ANALYSIS TERMS RELATED TO FOOD SAFETY[1]

Hazard: A biological, chemical or physical agent in, or condition of, food with the potential to cause an adverse health effect.

---

[1] These Definitions were adopted by the 22nd Session of the Commission (1997) on an interim basis: they are subject to modification in the light of developments in the science of risk analysis and as a result of efforts to harmonize similar definitions across various disciplines

Risk: A function of the probability of an adverse health effect and the severity of that effect, consequential to a hazard(s) in food.

Risk analysis: A process consisting of three components : risk assessment, risk management and risk communication.

Risk assessment: A scientifically based process consisting of the following steps: (i) hazard identification, (ii) hazard characterization, (iii) exposure assessment, and (iv) risk characterization.

Hazard identification: The identification of biological, chemical, and physical agents capable of causing adverse health effects and which may be present in a particular food or group of foods.

Hazard characterization: The qualitative and/or quantitative evaluation of the nature of the adverse health effects associated with biological, chemical and physical agents which may be present in food. For chemical agents, a dose-response assessment should be performed. For biological or physical agents, a dose-response assessment should be performed if the data are obtainable.

Dose-response assessment: The determination of the relationship between the magnitude of exposure (dose) to a chemical, biological or physical agent and the severity and/or frequency of associated adverse health effects (response).

Exposure assessment: The qualitative and/or quantitative evaluation of the likely intake of biological, chemical, and physical agents via food as well as exposures from other sources if relevant.

Risk characterization: The qualitative and/or quantitative estimation, including attendant uncertainties, of the probability of occurrence and severity of known or potential adverse health effects in a given population based on hazard identification, hazard characterization and exposure assessment.

Risk management: The process of weighing policy alternatives in the light of the results of risk assessment and, if required, selecting and implementing appropriate control options, including regulatory measures.

Risk communication: The interactive exchange of information and opinions concerning risk among risk assessors, risk managers, consumers and other interested parties.

SECTION II

Guidelines for Codex Committees
Guidelines for the Inclusion of Specific Provisions
Reference System for Documents
Format of Standards
Criteria for Work Priorities
Relations between Codex Committees

Contents of this Section

This Section of the Procedural Manual sets out the working procedures of the subsidiary bodies of the Codex Alimentarius Commission. It is primarily addressed to the Chairpersons and the Host Government Secretariats of individual Codex Committees.

The Guidelines for Codex Committees describe the organization and conduct of meetings and the preparation and distribution of working papers and reports. The Codex Reference System for Documents is explained in this Section.

Special technical sections are included on principles for the establishment or selection of Codex methods of analysis, Codex sampling procedures and on analytical terminology for use by Codex Committees.

The Format of Codex Standards and an explanatory note on how Committees should draft Codex Standards are described here. A section describes the Criteria for the Establishment of Work Priorities.

To ensure that the appropriate sections of Codex Commodity Standards have been reviewed for food safety, nutrition, consumer protection and food analysis, a section on the Relations between Commodity Committees and General Committees is included for guidance to Codex Commodity Committees.

GUIDELINES FOR CODEX COMMITTEES

INTRODUCTION

1. By virtue of Article 7 of the Statutes of the Codex Alimentarius Commission and Rule IX.1(b) of its Rules of Procedure, the Commission has established a number of Codex Committees to prepare standards in accordance with the Procedure for the Elaboration of Codex Standards and Coordinating Committees to exercise general coordination of its work in specific regions or groups of countries. The Rules of Procedure of the Commission shall apply, *mutatis mutandis*, to Codex Committees and Coordinating Committees.

COMPOSITION OF CODEX COMMITTEES

MEMBERSHIP

2. Membership of Codex Committees is open to Members of the Commission who have notified the Director-General of FAO or WHO of their desire to be considered as members thereof or to selected members designated by the Commission. Membership of Regional Coordinating Committees is open only to Members of the Commission belonging to the region or group of countries concerned.

OBSERVERS

3. Any other Member of the Commission or any Member or Associate Member of FAO or WHO which has not become a Member of the Commission may participate as an observer at any Codex Committee if it has notified the Director-General of FAO or WHO of its wish to do so. Such countries may participate fully in the discussions of the Committee and shall be provided with the same opportunities as other Members to express their point of view (including the submission of memoranda), but without the right to vote or to move motions either of substance or of procedure. International organizations which have formal relations with either FAO or WHO should also be invited to attend in an observer capacity sessions of those Codex Committees which are of interest to them.

ORGANIZATION AND DUTIES

CHAIR

4. The Codex Alimentarius Commission will designate a member country of the Commission, which has indicated its willingness to accept financial and all other responsibility, as having responsibility for appointing a chairperson of

the Committee. The member country concerned is responsible for appointing the chairperson of the Committee from among its own nationals. Should this person for any reason be unable to take the chair, the member country concerned shall designate another person to perform the functions of the chairperson for as long as the chairperson is unable to do so. A Committee may appoint at any session one or more rapporteurs from among the delegates present.

SECRETARIAT

5. A member country to which a Codex Committee has been assigned is responsible for providing all conference services including the secretariat. The secretariat should have adequate administrative support staff able to work easily in the languages used at the session and should have at its disposal adequate word processing and document reproducing equipment. Interpretation, preferably simultaneous, should be provided from and into all languages used at the session, and if the report of the session is to be adopted in more than one of the working languages of the Committee, then the services of a translator should be available. The Committee secretariat and the Joint FAO/WHO (Codex) Secretariat are charged with the preparation of the draft report in consultation with the rapporteurs, if any.

DUTIES AND TERMS OF REFERENCE

6. The duties of a Codex Committee shall include:

(a) the drawing up of a list of priorities as appropriate, among the subjects and products within its terms of reference,

(b) consideration of the types of safety and quality elements (or recommendations) to be covered, whether in standards for general application or in reference to specific food products,

(c) consideration of the types of product to be covered by standards, e.g., whether materials for further processing into food should be covered,

(d) preparation of draft Codex standards within its terms of reference,

(e) reporting to each session of the Commission on the progress of its work and, where necessary, on any difficulties caused by its terms of reference, together with suggestions for their amendment.

(f) the review and, as necessary, revision of existing standards and related texts on a scheduled, periodic basis to ensure that the standards and related texts within its terms of reference are consistent with current scientific knowledge and other relevant information.

SESSIONS

INVITATIONS AND PROVISIONAL AGENDA

7. (a) Sessions of Codex Committees and Coordinating Committees will be convened by the Directors-General of FAO and WHO in consultation with the chairperson of the respective Codex Committee. The letter of invitation and provisional agenda shall be prepared by the Chief, Joint FAO/WHO Food Standards Programme, FAO, Rome, in consultation with the chairperson of the committee for issue by the Directors-General to all Members and Associate Members of FAO and WHO or, in the case of Coordinating Committees, to the countries of the region or group of countries concerned, Codex Contact Points and interested international organizations in accordance with the official mailing lists of FAO and WHO. Chairpersons should, before finalizing the drafts, inform and consult with the national Codex Contact Point where one has been established, and, if necessary, obtain clearance from the national authorities concerned (Ministry of Foreign Affairs, Ministry of Agriculture, Ministry of Health, or as the case may be). The invitation and Provisional Agenda will be translated and distributed by FAO/WHO in the working languages of the Commission at least four months before the date of the meeting.

(b) Invitations should include the following:

(i) title of the Codex Committee,

(ii) time and date of opening and date of closing of the session,

(iii) place of the session,

(iv) languages to be used and arrangements for interpretation, i.e., whether simultaneous or not,

(v) if appropriate, information on hotel accommodation,

(vi) request for the names of the chief delegate and other members of the delegation, and for information on whether the chief delegate of a government will be attending as a representative or in the capacity of an observer.

Replies to invitations will normally be requested to be sent to reach the chairperson as early as possible and in any case not less than 30 days before the session. A copy should be sent also to the Chief, Joint FAO/WHO Food Standards Programme, FAO, Rome. It is of the utmost importance that by the date requested a reply to invitations should be sent by all those governments and international organizations which intend to participate. The reply should specify the number of copies and the language of the documents required.

(c) The Provisional Agenda should state the time, date and place of the meeting and should include the following items:

(i) adoption of the agenda,

(ii) if considered necessary, election of rapporteurs,

(iii) items relating to subject matter to be discussed, including, where appropriate, the step in the Commission's Procedure for the Elaboration of Standards at which the item is being dealt with at the session. There should also be reference to the Committee papers relevant to the item,

(iv) any other business,

(v) consideration of date and place of next session,

(vi) adoption of draft report.

The work of the Committee and the length of the meeting should be so arranged as to leave sufficient time at the end of the session for a report of the Committee's transactions to be agreed.

ORGANIZATION OF WORK

8. A Codex or Coordinating Committee may assign specific tasks to countries, groups of countries or to international organizations represented at meetings of the Committee and may ask member countries and international organizations for views on specific points. Ad hoc working groups established to accomplish specific tasks shall be disbanded once the tasks have been accomplished as determined by the Committee. Reports of ad hoc working groups shall be distributed to all members of the Committee and observers in time to allow full consideration of the working groups' recommendations. A Codex or Coordinating Committee may not set up standing sub-committees, whether open to all Members of the Commission or not, without the specific approval of the Commission.

PREPARATION AND DISTRIBUTION OF PAPERS

9. (a) Papers for a session should be sent by the chairperson of the Codex Committee concerned at least two months before the opening of the session to the following: (i) all Codex Contact Points, (ii) chief delegates of member countries, of observer countries and of international organizations, and (iii) other participants on the basis of replies received. Twenty copies of all papers in each of the languages used in the Committee concerned should be sent to the Chief, Joint FAO/WHO Food Standards Programme, FAO, Rome.

(b) Papers for a session prepared by participants must be drafted in one of the working languages of the Commission, which should, if possible, be one of the languages used in the Codex Committee concerned. These papers should be sent to the chairperson of the Committee, with a copy to the Chief, Joint FAO/WHO Food Standards Programme, FAO, Rome, in good time (see paragraph 9(a)) to be included in the distribution of papers for the session.

(c) Documents circulated at a session of a Codex Committee other than draft documents prepared at the session and ultimately issued in a final form, should subsequently receive the same distribution as other papers prepared for the Committee.

(d) Codex Contact Points will be responsible for ensuring that papers are circulated to those concerned within their own country and for ensuring that all necessary action is taken by the date specified.

(e) Consecutive reference numbers in suitable series should be assigned to all documents of Codex Committees. The reference number should appear at the top right-hand corner of the first page together with a statement of the language in which the document was prepared and the date of its preparation. A clear statement should be made of the provenance (origin or author country) of the paper immediately under the title. The text should be divided into numbered paragraphs. At the end of these guidelines is a series of references for Codex documents adopted by the Codex Alimentarius Commission for its own sessions and those of its subsidiary bodies.

(f) Members of the Codex Committees should advise the Committee chairperson through their Codex Contact Point of the number of copies should advise the Committee chairperson through their Codex Contact Point of the number of copies of documents normally required.

(g) Working papers of Codex Committees may be circulated freely to all those assisting a delegation in preparing for the business of the Committee; they should not, however, be published. There is, however, no objection to the publication of reports of the meetings of committees or of completed draft standards.

CONDUCT OF MEETINGS

10. (a) Meetings of Codex and Coordinating Committees shall be held in public unless the Committee decides otherwise. Member countries responsible for Codex and Coordinating Committees shall decide who should open meetings on their behalf. The chairperson should invite observations from members of the Committee concerning the Provisional Agenda and in the light of such observations formally request the

Committee to adopt the Provisional Agenda or the amended agenda. Meetings should be conducted in accordance with the Rules of Procedure of the Codex Alimentarius Commission. Attention is particularly drawn to Rule VI.7 which reads: "The provisions of Rule XII of the General Rules of FAO shall apply *mutatis mutandis* to all matters which are not specifically dealt with under Rule VI of the present Rules." Rule XII of the General Rules of FAO, a copy of which will be supplied to all chairpersons of Codex and Coordinating Committees, gives full instructions on the procedures to be followed in dealing with voting, points of order, adjournment and suspension of meetings, adjournment and closure of discussions on a particular item, reconsideration of a subject already decided and the order in which amendments should be dealt with.

(b) Chairpersons of Codex Committees should ensure that all questions are fully discussed, in particular statements concerning possible economic implications of standards under consideration at Steps 4 and 7. Chairpersons should also ensure that the written comments of members not present at the session are considered by the Committee; that all issues are put clearly to the Committee. This can usually best be done by stating what appears to be the generally acceptable view and asking delegates whether they have any objection to its being adopted. The chairpersons should always try to arrive at a consensus and should not ask the Committee to proceed to voting if agreement on the Committee's decision can be secured by consensus.

(c) Delegations and delegations from observer countries who wish their opposition to a decision of the Committee to be recorded may do so, whether the decision has been taken by a vote or not, by asking for a statement of their position to be contained in the report of the Committee. This statement should not merely use a phrase such as: "The delegation of X reserved its position" but should make clear the extent of the delegation's opposition to a particular decision of the Committee and state whether they were simply opposed to the decision or wished for a further opportunity to consider the question.

(d) Only the chief delegates of member countries, or of observer countries or of international organizations have the right to speak unless they authorize other members of their delegations to do so.

REPORTS

11. (a) In preparing reports, the following points shall be borne in mind:

    (i) decisions should be clearly stated; action taken in regard to economic impact statements should be fully recorded; all decisions on draft standards should be accompanied by an indication of the step in the Procedure that the standards have reached;

(ii) if action has to be taken before the next meeting of the committee, the nature of the action, who is to take it and when the action must be completed should be clearly stated;

(iii) where matters require attention by other Codex committees, this should be clearly stated;

(iv) if the report is of any length, summaries of points agreed and the action to be taken should be included at the end of the report, and in any case, a section should be included at the end of the report showing clearly in summary form:

(1) standards considered at the session and the steps they have reached;

(2) standards at any step of the Procedure, the consideration of which has been postponed or which are held in abeyance and the steps which they have reached;

(3) new standards proposed for consideration, the probable time of their consideration at Step 2 and the responsibility for drawing up the first draft.

(b) The following appendices should be attached to the report:

(i) list of participants with full postal addresses,

(ii) draft standards with an indication of the step in the Procedure which has been reached.

(c) The Joint FAO/WHO Secretariat should ensure that, as soon as posible and in any event not later than one month after the end of the session, copies of the final report, as adopted, are sent to all participants, and all Codex Contact Points.

DRAWING UP OF CODEX STANDARDS

12. A Codex committee, in drawing up standards and related texts, should bear in mind the following:

(a) The guidance given in the General Principles of the Codex Alimentarius;

(b) that all standards and related texts should have a preface containing the following information:

(i) the description of the standard or related text,

(ii) a brief description of the scope and purpose(s) of the standard or related text,

(iii) references including the step which the standard or related text has reached in the Commission's Procedures for the Elaboration of Standards, together with the date on which the draft was approved,

(iv) matters in the draft standard or related text requiring endorsement or action by other Codex Committees.

(c) that for standards or any related text for a product which includes a number of sub categories, the Committee should give preference to the development of a general standard or related text with specific provisions as necessary for sub-categories with different requirements.

GUIDELINES FOR THE INCLUSION OF SPECIFIC PROVISIONS IN CODEX STANDARDS AND RELATED TEXTS

GUIDELINES ON THE ELABORATION AND/OR REVISION OF CODES OF HYGIENIC PRACTICE FOR SPECIFIC COMMODITIES

The establishment of additional food hygiene requirements for specific food items or food groups should be limited to the extent necessary to meet the defined objectives of individual codes.

Codex Codes of Hygienic Practice should serve the primary purpose of providing advice to governments on the application of food hygiene provisions within the framework of national and international requirements.

The Revised Recommended International Code of Practice - General Principles of Food Hygiene (including the Guidelines for the Application of the Hazard Analysis Critical Control Point (HACCP) System) and the Revised Principles for the Establishment and Application of Microbiological Criteria for Foods are the base documents in the field of food hygiene.

All Codex Codes of Hygienic Practice applicable to specific food items or food groups shall refer to the General Principles of Food Hygiene and shall only contain material additional to the General Principles which is necessary to take into account the particular requirements of the specific food item or food group.

Provisions in Codex Codes of Hygienic Practice should be drafted in a sufficiently clear and transparent manner such that extended explanatory material is not required for their interpretation.

The above considerations should also apply to Codex Codes of Practice which contain provisions relating to food hygiene.

PRINCIPLES FOR THE ESTABLISHMENT OF CODEX METHODS OF ANALYSIS

PURPOSE OF CODEX METHODS OF ANALYSIS

The methods are primarily intended as international methods for the verification of provisions in Codex standards. They should be used for reference, in calibration of methods in use or introduced for routine examination and control purposes.

METHODS OF ANALYSIS

(A) Definition of types of methods of analysis

(a) Defining Methods (Type I)

Definition: A method which determines a value that can only be arrived at in terms of the method per se and serves by definition as the only method for establishing the accepted value of the item measured.

Examples: Howard Mould Count, Reichert-Meissl value, loss on drying, salt in brine by density.

(b) Reference Methods (Type II)

Definition: A Type II method is the one designated Reference Method where Type I methods do not apply. It should be selected from Type III methods (as defined below). It should be recommended for use in cases of dispute and for calibration purposes.

Example: Potentiometric method for halides.

(c) Alternative Approved Methods (Type III)

Definition: A Type III Method is one which meets the criteria required by the Codex Committee on Methods of Analysis and Sampling for methods that may be used for control, inspection or regulatory purposes.

Example: Volhard Method or Mohr Method for chlorides

(d) Tentative Method (Type IV)

Definition: A Type IV Method is a method which has been used traditionally or else has been recently introduced but for which the criteria required for acceptance by the Codex Committee on Methods of Analysis and Sampling have not yet been determined.

Examples: chlorine by X ray fluorescence, estimation of synthetic colours in foods.

(B) General Criteria for the Selection of Methods of Analysis

(a) Official methods of analysis elaborated by international organizations occupying themselves with a food or group of foods should be preferred.

(b) Preference should be given to methods of analysis the reliability of which have been established in respect of the following criteria, selected as appropriate:

    (i) specificity

(ii) accuracy

(iii) precision; repeatability intra-laboratory (within laboratory), reproducibility inter-laboratory (within laboratory and between laboratories)

(iv) limit of detection

(v) sensitivity

(vi) practicability and applicability under normal laboratory conditions

(vii) other criteria which may be selected as required.

(c) The method selected should be chosen on the basis of practicability and preference should be given to methods which have applicability for routine use.

(d) All proposed methods of analysis must have direct pertinence to the Codex Standard to which they are directed.

(e) Methods of analysis which are applicable uniformly to various groups of commodities should be given preference over methods which apply only to individual commodities.

ANALYTICAL TERMINOLOGY FOR CODEX USE

RESULT: The final value reported for a measured or computed quantity, after performing a measuring procedure including all subprocedures and evaluations.

NOTES:

1. When a result is given, it should be made clear whether it refers to:

    - the indication [signal]
    - the uncorrected result
    - the corrected result
    - and whether several values were averaged.

2. A complete statement of the result of a measurement includes information about the uncertainty of measurement.

SPECIFICITY: The property of a method to respond exclusively to the characteristic or analyte defined in the Codex standard.

NOTES:

1. Specificity may be achieved by many means: It may be inherent in the molecule (e.g., infrared or mass spectrometric identification techniques), or

attained by separations (e.g., chromatography), mathematically (e.g., simultaneous equations), or biochemically (e.g., enzyme reactions). Very frequently methods rely on the absence of interferences to achieve specificity (e.g., determination of chloride in the absence of bromide and iodide).

2. In some cases specificity is not desired (e.g., total fat, fatty acids, crude protein, dietary fibre, reducing sugars).

ACCURACY (AS A CONCEPT): The closeness of agreement between the reported result and the accepted reference value.

NOTE:

The term accuracy, when applied to a set of test results, involves a combination of random components and a common systematic error or bias component. When the systematic error component must be arrived at by a process that includes random error, the random error component is increased by propagation of error considerations and is reduced by replication.

ACCURACY (AS A STATISTIC): The closeness of agreement between a reported result and the accepted reference value.

NOTE:

Accuracy as a statistic applies to the single reported final test result; accuracy as a concept applies to single, replicate, or averaged values.

TRUENESS: The closeness of agreement between the average value obtained from a series of test results and an accepted reference value.

NOTES:

1. The measure of trueness is usually expressed in terms of bias.

2. Trueness has been referred to as "accuracy of the mean".

BIAS: The difference between the expectation of the test results and an accepted reference value.

NOTES:

1. Bias is the total systematic error as contrasted to random error. There may be one or more systematic error components contributing to bias. A larger systematic difference from the accepted reference value is reflected by a larger bias value.

2. When the systematic error component(s) must be arrived at by a process that includes random error, the random error component is increased by propagation of error considerations and reduced by replication.

PRECISION: The closeness of agreement between independent test results obtained under stipulated conditions.

NOTES:

1. Precision depends only on the distribution of random errors and does not relate to the true value or to the specified value.

2. The measure of precision is usually expressed in terms of imprecision and computed as a standard deviation of the test results. Less precision is reflected by a larger standard deviation.

3. "Independent test results" means results obtained in a manner not influenced by any previous result on the same or similar test object. Quantitative measures of precision depend critically on the stipulated conditions. Repeatability and reproducibility conditions are particular sets of extreme conditions.

   Repeatability [Reproducibility]: Precision under repeatability [reproducibility] conditions.

   Repeatability conditions: Conditions where independent test results are obtained with the same method on identical test items in the same laboratory by the same operator using the same equipment within short intervals of time.

   Reproducibility conditions: Conditions where test results are obtained with the same method on identical test items in different laboratories with different operators using different equipment.

   NOTE:

   When different methods give test results that do not differ significantly, or when different methods are permitted by the design of the experiment, as in a proficiency study or a material-certification study for the establishment of a consensus value of a reference material, the term "reproducibility" may be applied to the resulting parameters. The conditions must be explicitly stated.

   Repeatability [Reproducibility] standard deviation: The standard deviation of test results obtained under repeatability [reproducibility] conditions.

   NOTES:

   1. Repeatability [Reproducibility] standard deviation is a measure of the dispersion of the distribution of test results under repeatability [reproducibility] conditions.

2. Similarly "repeatability [reproducibility] variance" and "repeatability [reproducibility] coefficient of variation" could be defined and used as measures of the dispersion of test results under repeatability [reproducibility] conditions.

Repeatability [Reproducibility] limit: The value less than or equal to which the absolute difference between two test results obtained under repeatability [reproducibility] conditions may be expected to be with a probability of 95%.

NOTES:

1. The symbol used is r [R].

2. When examining two single test results obtained under repeatability [reproducibility] conditions, the comparison should be made with the repeatability [reproducibility] limit r [R] = 2.8 sr[sR].

3. When groups of measurements are used as the basis for the calculation of the repeatability [reproducibility] limits (now called the critical difference), more complicated formulae are required that are given in ISO 5725-6:1994, 4.2.1 and 4.2.2.

INTERLABORATORY STUDY: A study in which several laboratories measure a quantity in one or more "identical" portions of homogeneous, stable materials under documented conditions, the results of which are compiled into a single document.

NOTE:

The larger the number of participating laboratories, the greater the confidence that can be placed in the resulting estimates of the statistical parameters. The IUPAC-1987 protocol (Pure & Appl. Chem., 66, 1903-1911(1994)) requires a minimum of eight laboratories for method-performance studies.

Method-Performance Study: An interlaboratory study in which all laboratories follow the same written protocol and use the same test method to measure a quantity in sets of identical test samples. The reported results are used to estimate the performance characteristics of the method. Usually these characteristics are within-laboratory and among-laboratories precision, and when necessary and possible, other pertinent characteristics such as systematic error, recovery, internal quality control parameters, sensitivity, limit of determination, and applicability.

NOTES:

1. The materials used in such a study of analytical quantities are usually representative of materials to be analyzed in actual practice with respect to matrices, amount of test component (concentration), and interfering components and effects. Usually the analyst is not aware of the actual composition of the test samples but is aware of the matrix.
2. The number of laboratories, number of test samples, number of determinations, and other details of the study are specified in the study protocol. Part of the study protocol is the procedure which provides the written directions for performing the analysis.
3. The main distinguishing feature of this type of study is the necessity to follow the same written protocol and test method exactly.
4. Several methods may be compared using the same test materials. If all laboratories use the same set of directions for each method and if the statistical analysis is conducted separately for each method, the study is a set of method-performance studies. Such a study may also be designated as a method-comparison study.

Laboratory-Performance (Proficiency) Study: An interlaboratory study that consists of one or more measurements by a group of laboratories on one or more homogeneous, stable, test samples by the method selected or used by each laboratory. The reported results are compared with those from other laboratories or with the known or assigned reference value, usually with the objective of improving laboratory performance.

NOTES:
1. Laboratory-performance studies can be used to support accreditation of laboratories or to audit performance. If a study is conducted by an organization with some type of management control over the participating laboratories -- organizational, accreditation, regulatory, or contractual -- the method may be specified or the selection may be limited to a list of approved or equivalent methods. In such situations, a single test sample is insufficient to judge performance.
2. a laboratory-performance study may be used to select a method of analysis that will be used in a method-performance study. If all laboratories, or a sufficiently large subgroup, of laboratories, use the same method, the study may also be interpreted as a method-performance study, provided that the test samples cover the range of concentration of the analyte.

3. laboratories of a single organization with independent facilities, instruments, and calibration materials, are treated as different laboratories.

Material-Certification Study: An interlaboratory study that assigns a reference value ("true value") to a quantity (concentration or property) in the test material, usually with a stated uncertainty.

NOTE:

A material-certification study often utilizes selected reference laboratories to analyze a candidate reference material by a method(s) judged most likely to provide the least-biased estimates of concentration (or of a characteristic property) and the smallest associated uncertainty.

APPLICABILITY: The analytes, matrices, and concentrations for which a method of analysis may be used satisfactorily to determine compliance with a Codex standard.

NOTE:

In addition to a statement of the range of capability of satisfactory performance for each factor, the statement of applicability (scope) may also include warnings as to known interference by other analytes, or inapplicability to certain matrices and situations.

SENSITIVITY: Change in the response divided by the corresponding change in the concentration of a standard (calibration) curve; i.e., the slope, s_i, of the analytical calibration curve.

NOTE:

This term has been used for several other analytical applications, often referring to capability of detection, to the concentration giving 1% absorption in atomic absorption spectroscopy, and to ratio of found positives to known, true positives in immunological and microbiological tests. Such applications to analytical chemistry should be discouraged.

A method is said to be sensitive if a small change in concentration, c, or quantity, q, causes a large change in the measure, x; that is, when the derivative dx/dc or dx/dq is large.

Although the signal may vary with the magnitude of c_i or q_i, the slope, s_i, is usually constant over a reasonable range of concentrations. s_i may also be a function of the c or q of other analytes present in the sample.

RUGGEDNESS: The ability of a chemical measurement process to resist changes in results when subjected to minor changes in environmental and procedural variables, laboratories, personnel, etc.

PRINCIPLES FOR THE ESTABLISHMENT OR SELECTION OF CODEX SAMPLING PROCEDURES

PURPOSE OF CODEX METHODS OF SAMPLING

Codex Methods of Sampling are designed to ensure that fair and valid sampling procedures are used when food is being tested for compliance with a particular Codex commodity standard. The sampling methods are intended for use as international methods designed to avoid or remove difficulties which may be created by diverging legal, administrative and technical approaches to sampling and by diverging interpretation of results of analysis in relation to lots or consignments of foods, in the light of the relevant provision(s) of the applicable Codex standard.

METHODS OF SAMPLING

(A) Types of Sampling Plans and Procedures

(a) **Sampling Plans for Commodity Defects:**
These are normally applied to visual defects (e.g. loss of colour, mis-graded for size, etc.) and extraneous matter. They will normally be attributes plans, and plans such as those included in the *FAO/WHO Codex Alimentarius Sampling Plans for Prepackaged Foods (AQL 6.5)*[1] may be applied.

(b) **Sampling Plans for Net Contents:**
These are sampling plans which apply to pre-packaged foods generally and are intended to serve to check compliance of lots or consignments with provisions for net contents.

(c) **Sampling Plans for Compositional Criteria:**
Such plans are normally applied to analytically determined compositional criteria (e.g., loss on drying in white sugar, etc.). They are predominantly based on variable procedures with unknown standard deviation.

(d) **Specific Sampling Plans for Health-related Properties**

---

[1] *Codex Alimentarius*: Volume 13.

Such plans are generally applied to heterogeneous conditions, e.g., in the assessment of microbiological spoilage, microbial by-products or sporadically occurring chemical contaminants.

(B) General Instructions for the Selection of Methods of Sampling

(a) Official methods of sampling as elaborated by international organizations occupying themselves with a food or a group of foods are preferred. Such methods, when attracted to Codex standards, may be revised using Codex recommended sampling terms (to be elaborated).

(b) The appropriate Codex Commodity Committee should indicate, before it elaborates any sampling plan, or before any plan is endorsed by the Codex Committee on Methods of Analysis and Sampling, the following:

(i) the basis on which the criteria in the Codex Commodity standards have been drawn up (e.g. whether on the basis that every item in a lot, or a specified high proportion, shall comply with the provision in the standard or whether the average of a set of samples extracted from a lot must comply and, if so, whether a minimum or maximum tolerance, as appropriate, is to be given);

(ii) whether there is to be any differentiation in the relative importance of the criteria in the standards and, if so, what is the appropriate statistical parameter each criterion should attract, and hence, the basis for judgement when a lot is in conformity with a standard.

(c) Instructions on the procedure for the taking of samples should indicate the following:

(i) the measures necessary in order to ensure that the sample taken is representative of the consignment or of the lot;

(ii) the size and the number of individual items forming the sample taken from the lot or consignment;

(iii) the administrative measures for taking and handling the sample.

(d) The sampling protocol may include the following information:

(i) the statistical criteria to be used for acceptance or rejection of the lot on the basis of the sample;

(ii) the procedures to be adopted in cases of dispute.

GENERAL CONSIDERATIONS

(a) The Codex Committee on Methods of Analysis and Sampling should maintain closest possible relations with all interested organizations working on methods of analysis and sampling.

(b) The Codex Committee on Methods of Analysis and Sampling should organize its work in such a manner as to keep under constant review all methods of analysis and sampling published in the Codex Alimentarius.

(c) In the Codex methods of analysis, provision should be made for variations in reagent concentrations and specifications from country to country.

(d) Codex methods of analysis which have been derived from scientific journals, theses, or publications, either not readily available or available in languages other than the official languages of FAO and WHO, or which for other reasons should be printed in the Codex Alimentarius *in extenso*, should follow the standard layout for methods of analysis as adopted by the Codex Committee on Methods of Analysis and Sampling.

(e) Methods of analysis which have already been printed as official methods of analysis in other available publications and which are adopted as Codex methods need only be quoted by reference in the Codex Alimentarius.

UNIFORM SYSTEM OF REFERENCES FOR CODEX DOCUMENTS

In referencing Codex documents, CX, which stands for Codex, should appear first, followed by the subject code reference, followed by the year in which the session will be held (i.e. not necessarily the year in which the document is prepared), and finally followed by the consecutive number of the document.

For example documents prepared for a session of the Codex Regional Coordinating Committee for Africa, meeting in 1994, would be identified by the series CX/AFRICA 94/1, 2, 3 etc. The only exception is the Executive Committee in which the session number is also identified: for example CX/EXEC 94/41/1, 2, 3 etc.

Codex Alimentarius Commission
(working documents and reports) .... ALINORM

Executive Committee
(identified also by session number following the year) .... CX/EXEC

Regional Coordinating Committees

Coordinating Committee for Africa .... CX/AFRICA

Coordinating Committee for Asia .... CX/ASIA

Coordinating Committee for Europe .... CX/EURO

Coordinating Committee for Latin America and the Caribbean .... CX/LAC

Coordinating Committee for North America
and the South West Pacific .... CX/NASWP

Codex Committees

General Principles .... CX/GP

Food Additives and Contaminants .... CX/FAC

Food Hygiene .... CX/FH

Food Labelling .... CX/FL

Methods of Analysis and Sampling .... CX/MAS

Pesticide Residues .... CX/PR

Residues of Veterinary Drugs in Foods .... CX/RVDF

Food Import and Export Inspection and Certification Systems .... CX/FICS
Nutrition and Foods for Special Dietary Uses .... CX/NFSDU
Cereals, Pulses and Legumes .... CX/CPL
Cocoa Products and Chocolate .... CX/CPC
Fats and Oils .... CX/FO
Fish and Fishery Products .... CX/FFP
Milk and Milk Products .... CX/MMP
Meat Hygiene .... CX/MH
Natural Mineral Waters .... CX/MIN
Processed Fruits and Vegetables .... CX/PFV
Processed Meat and Poultry Products .... CX/PMPP
Soups and Broths .... CX/SB
Sugars .... CX/S
Vegetable Proteins .... CX/VP
Fresh Fruits and Vegetables .... CX/FFV

ECE/Codex Alimentarius Groups of Experts

Fruit Juices .... CX/FJ
Quick Frozen Foods .... CX/QFF

… Procedural Manual …

FORMAT FOR CODEX COMMODITY STANDARDS INCLUDING STANDARDS ELABORATED UNDER THE CODE OF PRINCIPLES CONCERNING MILK AND MILK PRODUCTS

INTRODUCTION

The Format is also intended for use as a guide by the subsidiary bodies of the Codex Alimentarius Commission in presenting their standards, with the object of achieving, as far as possible, a uniform presentation of commodity standards. The Format also indicates the statements which should be included in standards as appropriate under the relevant headings of the standard. The sections of the Format require to be completed in a standard only insofar as such provisions are appropriate to an international standard for the food in question.

NAME OF THE STANDARD

SCOPE

DESCRIPTION

ESSENTIAL COMPOSITION AND QUALITY FACTORS

FOOD ADDITIVES

CONTAMINANTS

HYGIENE

WEIGHTS AND MEASURES

LABELLING

METHODS OF ANALYSIS AND SAMPLING

NOTES ON THE HEADINGS

NAME OF THE STANDARD

The name of the standard should be clear and as concise as possible. It should usually be the common name by which the food covered by the standard is known or, if more than one food is dealt with in the standard, by a generic name covering them all. If a fully informative title should be inordinately long, a subtitle could be added.

SCOPE

This section should contain a clear, concise statement as to the food or foods to which the standard is applicable unless this is self explanatory in the name of the standard. In the case of a general standard covering more than one specific product, it should be made clear as to which specific products the standard applies.

DESCRIPTION

This section should contain a definition of the product or products with an indication, where appropriate, of the raw materials from which it is derived and any necessary references to processes of manufacture. It may also include references to types and styles of product and to type of pack. There may also be additional definitions when these are required to clarify the meaning of the standard.

ESSENTIAL COMPOSITION AND QUALITY FACTORS

This section should contain all quantitative and other requirements as to composition including, where necessary, identity characteristics, provisions on packing media and requirements as to compulsory and optional ingredients. It should also include quality factors which are essential for the designation, definition or composition of the product concerned. Such factors could include the quality of the raw material, with the object of protecting the health of the consumer, provisions on taste, odour, colour and texture which may be apprehended by the senses, and basic quality criteria for the finished products, with the object of preventing fraud. This section may refer to tolerances for defects, such as blemishes or imperfect material, but this information should be contained in an appendix to the standard or in another advisory text.

FOOD ADDITIVES

This section should contain the names of the additives permitted and, where appropriate, the maximum amount permitted in the food. It should be prepared in accordance with guidance given on page 76 and may take the following form:

> "The following provisions in respect of food additives and their specifications as contained in section .... of the Codex Alimentarius are subject to endorsement [have been endorsed] by the Codex Committee on Food Additives and Contaminants."

Then should follow a tabulation, viz.:

> "Name of additive, maximum level (in percentage or mg/kg)."

CONTAMINANTS

(a) Pesticide Residues: This section should include, by reference, any levels for pesticide residues that have been established by the Codex Alimentarius Commission for the product concerned.[1]

(b) Other Contaminants: In addition, this section should contain the names of other contaminants and where appropriate the maximum level permitted in the food, and the text to appear in the standard may take the following form:

"The following provisions in respect of contaminants, other than pesticide residues, are subject to endorsement [have been endorsed] by the Codex Committee on Food Additives and Contaminants."

Then should follow a tabulation, viz.:

"Name of contaminant, maximum level (in percentage or mg/kg)."

HYGIENE

Any specific mandatory hygiene provisions considered necessary should be included in this section. They should be prepared in accordance with the guidance given on page 78. Reference should also be made to applicable codes of hygienic practice. Any parts of such codes, including in particular any end-product specifications, should be set out in the standard, if it is considered necessary that they should be made mandatory. The following statement should also appear:

"The following provisions in respect of the food hygiene of this product are subject to endorsement [have been endorsed] by the Codex Committee on Food Hygiene."

WEIGHTS AND MEASURES

This section should include all provisions, other than labelling provisions, relating to weights and measures, e.g. where appropriate, fill of container, weight, measure or count of units determined by an appropriate method of sampling and analysis. Weights and measures should be expressed in S.I. units. In the case of standards which include provisions for the sale of products in standardized amounts, e.g. multiples of 100 grams, S.I. units should be used, but this would not preclude additional statements in the

---

[1] N.B. This Procedure has not been followed for practical reasons. Codex maximum limits for pesticide residues are published separately in Volume 2 of the *Codex Alimentarius*.

standards of these standardized amounts in approximately similar amounts in other systems of weights and measures.

LABELLING

This section should include all the labelling provisions contained in the standard and should be prepared in accordance with the guidance given on page 75. Provisions should be included by reference to the General Standard for the Labelling of Prepackaged Foods. The section may also contain provisions which are exemptions from, additions to, or which are necessary for the interpretation of the General Standard in respect of the product concerned provided that these can be justified fully. The following statement should also appear:

> *"The following provisions in respect of the labelling of this product are subject to endorsement [have been endorsed] by the Codex Committee on Food Labelling."*

METHODS OF ANALYSIS AND SAMPLING

This section should include, either specifically or by reference, all methods of analysis and sampling considered necessary and should be prepared in accordance with the guidance given on page 79. If two or more methods have been proved to be equivalent by the Codex Committee on Methods of Analysis and Sampling, these could be regarded as alternative and included in this section either specifically or by reference. The following statement should also appear:

> *"The methods of analysis and sampling described hereunder are to be endorsed [have been endorsed] by the Codex Committee on Methods of Analysis and Sampling."* [1]

---

[1] Methods of analysis should be indicated as being "defining", "reference", "alternative approved" or "tentative" methods, as appropriate. See page 56.

CRITERIA FOR THE ESTABLISHMENT OF WORK PRIORITIES AND FOR THE ESTABLISHMENT OF SUBSIDIARY BODIES OF THE CODEX ALIMENTARIUS COMMISSION

NEW WORK TO BE UNDERTAKEN BY EXISTING SUBSIDIARY BODIES

1. When a Codex Committee proposes to elaborate a standard, code of practice or related text within its terms of reference, it should first consider the priorities established by the Commission in the Medium-Term Plan of Work, any specific relevant strategic project currently being undertaken by the Commission and the prospect of completing the work within a reasonable period of time. It should also assess the proposal against the criteria set out in paragraph 4, below.

2. If the proposal falls in an area outside the Committee's terms of reference the proposal should be reported to the Commission in writing together with proposals for such amendments to the Committee's terms of reference as may be required.

NEW WORK WHICH WOULD REQUIRE THE ESTABLISHMENT OF A NEW SUBSIDIARY BODY

3. When a Member wishes to propose the elaboration of a standard, code of practice or related text in an area not covered by the terms of reference of any existing subsidiary body, it should accompany its proposal with a written statement to the Commission referring to the Commission's Medium-Term Objectives and containing, as far as practicable, the information required by the appropriate section of paragraph 4, below.

4. CRITERIA

A. Criteria applicable to commodities

(i) Consumer protection from the point of view of health and fraudulent practices.

(ii) Volume of production and consumption in individual countries and volume and pattern of trade between countries.

(iii) Diversification of national legislations and apparent resultant impediments to international trade.

(iv) International or regional market potential.

(v) Amenability of the commodity to standardization.

(vi) Number of commodities which would need separate standards indicating whether raw, semi processed or processed.

(vii) Work already undertaken by other international organizations in this field.

(viii) The type of subsidiary body envisaged to undertake the work.

B. ***Criteria applicable to general subjects***

(i) Consumer protection from the point of view of health and fraudulent practices.

(ii) Diversification of national legislations and apparent resultant impediments to international trade.

(iii) Scope of work and establishment of priorities between the various sections of the work.

(iv) Work already undertaken by other international organizations in this field.

(v) Type of subsidiary body envisaged to undertake the work.

RELATIONS BETWEEN COMMODITY COMMITTEES AND GENERAL COMMITTEES

Codex Committees may ask the advice and guidance of committees having responsibility for matters applicable to all foods on any points coming within their province.

The Codex Committees on Food Labelling; Food Additives and Contaminants; Methods of Analysis and Sampling; Food Hygiene; Nutrition and Foods for Special Dietary Uses; and Food Import and Export Inspection and Certification Systems may establish general provisions on matters within their terms of reference. These provisions should only be incorporated into Codex Commodity Standards by reference unless there is a need for doing otherwise.

Codex Commodity standards shall contain sections on hygiene, labelling and methods of analysis and sampling and these sections should contain all of the relevant provisions of the standard. Provisions of Codex General Standards, Codes or Guidelines shall only be incorporated into Codex Commodity Standards by reference unless there is a need for doing otherwise. Where Codex Committees are of the opinion that the general provisions are not applicable to one or more commodity standards, they may request the responsible Committees to endorse deviations from the general provisions of the Codex Alimentarius. Such requests should be fully justified and supported by available scientific evidence and other relevant information. Sections on hygiene, labelling, and methods of analysis and sampling which contain specific provisions or provisions supplementing the Codex General Standards, Codes or Guidelines shall be referred to the responsible Codex Committees at the most suitable time during Steps 3, 4 and 5 of the Procedure for the Elaboration of Codex Standards and Related Texts, though such reference should not be allowed to delay the progress of the standard to the subsequent steps of the Procedure.

Subject and commodity Committees should refer to the principles and guidelines developed by the Codex Committee on Food Import and Export Inspection and Certification Systems when developing provisions and/or recommendations on inspection and certification and make any appropriate amendments to the standards, guidelines and codes within the responsibility of the individual committees at the earliest convenient time.

FOOD LABELLING

The provisions on food labelling should be included by reference to the Codex General Standard for the Labelling of Prepackaged Foods (CODEX STAN 1-

1985). Exemptions from, or additions to, the General Standard which are necessary for its interpretation in respect of the product concerned should be justified fully, and should be restricted as much as possible.

Information specified in each draft standard should normally be limited to the following:

- a statement that the product shall be labelled in accordance with the Codex General Standard for the Labelling of Prepackaged Foods (CODEX STAN 1-1985)
- the specified name of the food
- date marking and storage instructions (only if the exemption foreseen in Section 4.7.1 of the General Standard is applied)

Where the scope of the Codex Standard is not limited to prepackaged foods, a provision for labelling of non retail containers may be included.

In such cases the provision may specify that "Information on ..."[1] shall be given either on the container or in accompanying documents, except that the name of the product, lot identification, and the name and address of the manufacturer or packer shall appear on the container.[2]

However, lot identification, and the name and address of the manufacturer or packer may be replaced by an identification mark provided that such a mark is clearly identifiable with the accompanying documents."

In respect of date marking (Section 4.7 of the General Standard), a Codex Committee may, in exceptional circumstances, determine another date or dates as defined in the General Standard, either to replace or to accompany the date of minimum durability, or alternatively decide that no date marking is necessary. In such cases, a full justification for the proposed action should be submitted to the Codex Committee on Food Labelling.

FOOD ADDITIVES AND CONTAMINANTS

Codex commodity committees should prepare a section on food additives in each draft commodity standard and this section should contain all the provisions in the standard relating to food additives. The section should include the names of those additives which are considered to be technologically necessary or

---

[1] Codex Committees should decide which provisons are to be included.

[2] Codex Committees may decide that further information is required on the container. In this regard, special attention should be given to the need for storage instructions to be included on the container.

which are widely permitted for use in the food within maximum levels where appropriate.

All provisions in respect of food additives (including processing aids) and contaminants contained in Codex commodity standards should be referred to the Codex Committee on Food Additives and Contaminants preferably after the Standards have been advanced to Step 5 of the Procedure for the Elaboration of Codex Standards or before they are considered by the Commodity Committee concerned at Step 7, though such reference should not be allowed to delay the progress of the Standard to the subsequent Steps of the Procedure.

All provisions in respect of food additives will require to be endorsed by the Codex Committee on Food Additives and Contaminants, on the basis of technological justification submitted by the commodity committees and of the recommendations of the Joint FAO/WHO Expert Committee on Food Additives concerning the safety-in-use (acceptable daily intake (ADI) and other restrictions) and an estimate of the potential and, where possible, the actual intake of the food additives, ensuring conformity with the General Principles for the Use of Food Additives.

In preparing working papers for the Codex Committee on Food Additives, the Secretariat should make a report to the Committee concerning the endorsement of provisions for food additives (including processing aids), on the basis of the General Principles for the Use of Food Additives. Provisions for food additives should indicate the International Numbering System (INS) number, the ADI, technological justification, proposed level, and whether the additive was previously endorsed (or temporarily endorsed).

When commodity standards are sent to governments for comment at Step 3, they should contain a statement that the provisions "in respect of food additives and contaminants are subject to endorsement by the Codex Committee on Food Additives and Contaminants and to and to incorporation into the General Standard for Food Addtives or the General Standard for Contamiants and Toxins in Foods."

When establishing provisions for food additives, Codex committees should follow the General Principles for the Use of Food Additives and the Preamble of the General Standard for Food Additives. Full explanation should be provided for any departure from the above recommendations.

When an active commodity committee exists, proposals for the use of additives in any commodity standard under consideration should be prepared by the committee concerned, and forwarded to the Codex Committee on Food Additives and Contaminants for endorsement. When the Codex Committee on Food Additives and Contaminants decides not to endorse specific additives provisions (use of the additive, or level in the end-product), the reason should be clearly stated. The section under consideration should be referred back to the

Committee concerned if further information is needed, or for information if the Codex Committee on Food Additives and Contaminants decides to amend the provision.

When no active commodity committee exists, proposals for new additive provisions or amendment of existing provisions, should be forwarded directly by member countries to the Codex Committee on Food Additives and Contaminants.

Good Manufacturing Practice means that:

(i) the quantity of the additive added to food does not exceed the amount reasonably required to accomplish its intended physical nutritional or other technical effect in food;

(ii) the quantity of the additive that becomes a component of food as a result of its use in the manufacturing, processing or packaging of a food and which is not intended to accomplish any physical, or other technological effect in the food itself, is reduced to the extent reasonably possible;

(iii) the additive is of appropriate food grade quality and is prepared and handled in the same way as a food ingredient. Food grade quality is achieved by compliance with the specifications as a whole and not merely with individual criteria in terms of safety.

FOOD HYGIENE

Commodity Committees may wish to select one of the following texts according to the nature of the product subject of the standard:

(i) For shelf stable products where microbiological spoilage before or after process is unlikely to be of significance:

It is recommended that the product covered by the provisions of this Standard be prepared in accordance with the appropriate sections of the General Principles of Food Hygiene recommended by the Codex Alimentarius Commission (Ref. No. CAC/RCP 1 1969, Rev. 3, 1997).

(ii) For shelf stable products, heat processed in hermetically sealed containers:

It is recommended that the product covered by the provision of this standard be prepared in accordance with the General Principles of Food Hygiene (CAC/RCP 1 1969, Rev. 3, 1997) and, where appropriate, with the Code of Hygienic Practice for Low Acid and Acidified Low Acid Canned Foods (Ref. CAC/RCP 23 1979, Rev. 1, 1989) or other Codes of Hygienic Practice as recommended by the Codex Alimentarius Commission.

To the extent possible in good manufacturing practice, the product shall be free from objectionable matter.

When tested by appropriate methods of sampling and examination, the product:

(a) shall be free from microorganisms capable of development in the food under normal conditions of storage; and

(b) shall not contain any substance originating from microorganisms in amounts which may represent a health hazard.

(iii) For all other products:

It is recommended that the product covered by the provisions of this standard be prepared and handled in accordance with the appropriate sections of the Recommended International Code of Practice General Principles of Food Hygiene (CAC/RCP 1 1969, Rev.3, 1997), and other Codes of Practice recommended by the Codex Alimentarius Commission which are relevant to this product. (A list may follow).

To the extent possible in good manufacturing practice, the product shall be free from objectionable matter.

When tested by appropriate methods of sampling and examination, the product:

(a) shall be free from microorganisms in amounts which may represent a hazard to health;

(b) shall be free from parasites which may represent a hazard to health; and

(c) shall not contain any substance originating from microorganisms in amounts which may represent a hazard to health.

METHODS OF ANALYSIS AND SAMPLING

NORMAL PRACTICE

Except for methods of analysis and sampling associated with microbiological criteria, when Codex committees have included provisions on methods of analysis or sampling in a Codex commodity standard, these should be referred to the Codex Committee on Methods of Analysis and Sampling at Step 4, to ensure Government comments at the earliest possible stage in the development of the standard. A Codex Committee should, whenever possible, provide to the Codex Committee on Methods of Analysis and Sampling information, for each individual analytical method proposed, relating to specificity, accuracy, precision (repeatability, reproducibility) limit of detection, sensitivity, applicability and practicability, as appropriate. Similarly a Codex Committee should, whenever possible, provide to the Codex Committee on Methods of Analysis and Sampling information for each sampling plan relating to the scope or field of application, the type of sampling (e.g. bulk or unit), sample sizes, decision rules, details of plans (e.g. "Operating characteristic" curves), inferences to be made to lots or processes, levels of risk to be accepted and pertinent supportive data.

Other criteria may be selected as required. Methods of analysis should be proposed by the Commodity Committees in consultation if necessary with an expert body.

At Step 4 Codex Commodity Committees should discuss and report to the Codex Committee on Methods of Analysis and Sampling on matters connected with:

- Provisions in Codex standards which require analytical or statistical procedure;
- Provisions for which elaboration of specific methods of analysis or sampling are required;
- Provisions which are defined by the use of Defining Methods (Type I);
- All proposals to the extent possible should be supported by appropriate documentation; especially for Tentative Methods (Type IV);
- Any request for advice or assistance.

The Codex Committee on Methods of Analysis and Sampling should undertake a coordinating role in matters relating to the elaboration of Codex methods of analysis and sampling. The originating committee is, however, responsible for carrying out the Steps of the Procedure.

When it is necessary, the Codex Committee on Methods of Analysis and Sampling should try to ensure elaboration and collaborative testing of methods by other recognized bodies with expertise in the field of analysis.

METHODS OF ANALYSIS AND SAMPLING OF GENERAL APPLICATION TO FOODS

When the Codex Committee on Methods of Analysis and Sampling itself elaborates methods of analysis and sampling which are of general application to foods, it is responsible for carrying out the steps of the Procedure.

METHODS OF ANALYSIS OF FOOD ADDITIVES AS SUCH

Methods of analysis included in Codex Advisory Food Additives Specifications, for the purpose of verifying the criteria of purity and identity of the food additive, need not be referred to the Codex Committee on Methods of Analysis and Sampling for endorsement. The Codex Committee on Food Additives and Contaminants is responsible for carrying out the steps of the Procedure.

METHODS OF ANALYSIS OF PESTICIDE RESIDUES IN FOOD

The methods for determining the levels of pesticide residues in food need not be referred to the Codex Committee on Methods of Analysis and Sampling for endorsement. The Codex Committee on Pesticide Residues is responsible for carrying out the steps of the Procedure.

MICROBIOLOGICAL METHODS OF ANALYSIS AND SAMPLING

When Codex committees have included provisions on microbiological methods of analysis and sampling for the purpose of verifying hygiene provisions, they should be referred to the Codex Committee on Food Hygiene at the most suitable time during Steps 3, 4 and 5 of the Procedure for the Elaboration of Codex Standards, which will ensure that government comments on the methods of analysis and sampling arc available to the Codex Committee on Food Hygiene. The procedure to be followed will be as in the normal practice described above, substituting the Codex Committee on Food Hygiene for the Codex Committee on Methods of Analysis and Sampling. Microbiological methods of analysis and sampling elaborated by the Codex Committee on Food Hygiene for inclusion in Codex commodity standards for the purpose of verifying hygiene provisions need not be referred to the Codex Committee on Methods of Analysis and Sampling for endorsement.

QUICK FROZEN FOODS

When Codex committees have elaborated Codex commodity standards for quick frozen food products, these should be referred to the Joint ECE/Codex Alimentarius Group of Experts on Standardization of Quick Frozen Foods at the most suitable time during Steps 3, 4 and 5 of the Procedure for the Elaboration of Codex Standards for comment by the group of experts.

SECTION III
Subsidiary Bodies
Membership
Contact Points
Organigram

Contents of this Section

This Section contains factual information about the Codex Alimentarius Commission, including a list of the Commission's Sessions.

The list of the Commission's Subsidiary Bodies gives the Terms of Reference of all Codex Committees established under Rule IX.1 of the Commission's Rules of Procedure. The meetings of each Committee are listed. The structure of the Commission's subsidiary bodies is shown diagrammatically on the inside back cover.

The countries which form the Commission's Membership are listed (as of July 1997) together with a list of the national Codex Contact Points. These lists are subject to frequent changes. The Secretariat of the Joint FAO/WHO Food Standards Programme provides up-dated information at regular intervals.

SESSIONS OF THE CODEX ALIMENTARIUS COMMISSION

1st,	Rome, Italy, 25 June - 3 July 1963
2nd,	Geneva, Switzerland, 28 September - 7 October 1964
3rd,	Rome, Italy, 19-28 October 1965
4th,	Rome, Italy, 7-14 November 1966
5th,	Rome, Italy, 20 February - 1 March 1968
6th,	Geneva, Switzerland, 4-14 March 1969
7th,	Rome, Italy, 7-17 April 1970
8th,	Geneva, Switzerland, 30 June - 9 July 1971
9th,	Rome, Italy, 6-17 November 1972
10th,	Rome, Italy, 1-11 July 1974
11th,	Rome, Italy, 29 March - 9 April 1976
12th,	Rome, Italy, 17-28 April 1978
13th,	Rome, Italy, 3-14 December 1979
14th,	Geneva, Switzerland, 29 June - 10 July 1981
15th,	Rome, Italy 4-15 July 1983
16th,	Geneva Switzerland, 1-12 July 1985
17th,	Rome, Italy, 29 June - 10 July 1987
18th,	Geneva, Switzerland, 3-12 July 1989
19th,	Rome, Italy, 1-10 July 1991
20th,	Geneva, Switzerland, 28 June - 7 July 1993
21st,	Rome, Italy, 3-8 July 1995
22nd,	Geneva, Switzerland, 23-28 June 1997

LIST OF SUBSIDIARY BODIES OF THE CODEX ALIMENTARIUS COMMISSION

SUBSIDIARY BODY UNDER RULE IX.1(A)

JOINT FAO/WHO COMMITTEE OF GOVERNMENT EXPERTS ON THE CODE OF PRINCIPLES CONCERNING MILK AND MILK PRODUCTS

Established by FAO and WHO in 1958 and integrated into the Joint FAO/WHO Food Standards Programme in 1962 as a subsidiary body of the Codex Alimentarius Commission under Rule IX.1(a). Re-named "Codex Committee on Milk and Milk Products" in 1993 and re-established as a subsidiary body under Rule IX.1(b)(i) see page 104.

Sessions

1st,	Rome, Italy, 8-12 September 1958
2nd,	Rome, Italy, 13-17 April 1959
3rd,	Rome, Italy, 22-26 February 1960
4th,	Rome, Italy, 6-10 March 1961
5th,	Rome, Italy, 2-6 April 1962
6th,	Rome, Italy, 17-21 June 1963
7th,	Rome, Italy, 4-8 May 1964
8th,	Rome, Italy, 24-29 May 1965
9th,	Rome, Italy, 20-25 June 1966
10th,	Rome, Italy, 25-31 August 1967
11th,	Rome, Italy, 10-15 June 1968
12th,	Rome, Italy, 7-12 July 1969
13th,	Rome, Italy, 15-20 June 1970
14th,	Rome, Italy, 6-11 September 1971
15th,	Rome, Italy, 25-30 September 1972
16th,	Rome, Italy, 10-15 September 1973
17th,	Rome, Italy, 14-19 April 1975
18th,	Rome, Italy, 13-18 September 1976
19th,	Rome, Italy, 12-17 June 1978
20th,	Rome, Italy, 26-30 April 1982
21st,	Rome, Italy, 2-6 June 1986
22nd,	Rome, Italy, 5-9 November 1990

Terms of Reference:

To establish international codes and standards concerning milk and milk products.

SUBSIDIARY BODIES UNDER RULE IX.1(B)(I)

CODEX COMMITTEE ON GENERAL PRINCIPLES

Host Government: France

Sessions:

1st,	Paris, 4-8 October 1965
2nd,	Paris, 16-19 October 1967
3rd,	Paris, 9-13 December 1968
4th,	Paris, 4-8 March 1974
5th,	Paris, 19-23 January 1976
6th,	Paris, 15-19 October 1979
7th,	Paris, 6-10 April 1981
8th,	Paris, 24-28 November 1986
9th,	Paris, 24-28 April 1989
10th,	Paris, 7-11 September 1992
11th,	Paris, 25-29 April 1994
12th,	Paris, 25-28 November 1996

Terms of Reference:

To deal with such procedural and general matters as are referred to it by the Codex Alimentarius Commission. Such matters have included the establishment of the General Principles which define the purpose and scope of the Codex Alimentarius, the nature of Codex standards and the forms of acceptance by countries of Codex standards; the development of Guidelines for Codex Committees; the development of a mechanism for examining any economic impact statements submitted by governments concerning possible implications for their economies of some of the individual standards or some of the provisions thereof; the establishment of a Code of Ethics for the International Trade in Food.

CODEX COMMITTEE ON FOOD ADDITIVES AND CONTAMINANTS

Host Government: Netherlands

Sessions:

1st,	The Hague, 19-22 May 1964
2nd,	The Hague, 10-14 May 1965
3rd,	The Hague, 9-13 May 1966
4th,	The Hague, 11-15 September 1967
5th,	Arnhem, 18-22 March 1968
6th,	Arnhem, 15-22 October 1969
7th,	The Hague, 12-16 October, 1970
8th,	Wageningen, 29 May - 2 June 1972
9th,	Wageningen, 10-14 December 1973
10th,	The Hague, 2-7 June 1975
11th,	The Hague, 31 May - 6 June 1977
12th,	The Hague, 10-16 October 1978
13th,	The Hague, 11-17 September 1979
14th,	The Hague, 25 Nov. - 1 Dec. 1980
15th,	The Hague, 16-22 March 1982
16th,	The Hague, 22-28 March 1983
17th,	The Hague, 10-16 April 1984
18th,	The Hague, 5-11 November 1985
19th,	The Hague, 17-23 March 1987
20th,	The Hague, 7-12 March 1988
21st,	The Hague, 13-18 March 1989
22nd,	The Hague, 19-24 March 1990
23rd,	The Hague, 4-9 March 1991
24th,	The Hague, 23-28 March 1992
25th,	The Hague, 22-26 March 1993
26th,	The Hague, 7-11 March 1994
27th,	The Hague, 20-24 March 1995
28th,	Manila, 18-22 March 1996
29th,	The Hague, 17-21 March 1997

Terms of reference:

(a) to establish or endorse permitted maximum or guideline levels for individual food additives, for contaminants (including environmental contaminants) and for naturally occurring toxicants in foodstuffs and animal feeds;

(b) to prepare priority lists of food additives and contaminants for toxicological evaluation by the Joint FAO/WHO Expert Committee on Food Additives;

(c) to recommend specifications of identity and purity for food additives for adoption by the Commission;

(d) to consider methods of analysis for their determination in food; and

(e) consider and elaborate standards or codes for related subjects such as the labelling of food additives when sold as such, and food irradiation.

CODEX COMMITTEE ON FOOD HYGIENE

Host Government: U.S.A.

Sessions:

1st,	Washington D.C., 27-28 May 1964
2nd,	Rome, 14-16 June 1965
3rd,	Rome, 31 May - 3 June 1966
4th,	Washington D.C., 12-16 June 1967
5th,	Washington D.C., 6-10 May 1968
6th,	Washington D.C., 5-9 May 1969
7th,	Washington D.C., 25-29 May 1970
8th,	Washington D.C., 14-18 June 1971
9th,	Washington D.C., 19-23 June 1972
10th,	Washington D.C., 14-18 May 1973
11th,	Washington D.C., 10-14 June 1974
12th,	Washington D.C., 12-16 May 1975
13th,	Rome, 10-14 May 1976
14th,	Washington D.C., 29 August - 2 September 1977
15th,	Washington D.C., 18-22 September 1978
16th,	Washington D.C., 23-27 July 1979
17th,	Washington D.C., 17-21 November 1980
18th,	Washington D.C., 22-26 February 1982
19th,	Washington D.C., 26-30 September 1983
20th,	Washington D.C., 1-5 October 1984
21st,	Washington D.C., 23-27 September 1985
22nd,	Washington D.C., 20-24 October 1986
23rd,	Washington D.C., 21-25 March 1988
24th,	Washington D.C., 16-20 October 1989
25th,	Washington D.C., 28 October - 1 November 1991
26th,	Washington D.C., 1-5 March 1993

Procedural Manual Page 89

27th, Washington D.C., 17-21 October 1994
28th, Washington D.C., 27 November - 1 December 1995
29th, Washington D.C., 21-25 October 1996

Terms of reference:

(a) to draft basic provisions on food hygiene applicable to all food[1];

(b) (i) to consider, amend if necessary and endorse provisions on hygiene prepared by Codex commodity committees and contained in Codex commodity standards, and

(ii) to consider, amend if necessary, and endorse provisions on hygiene prepared by Codex commodity committees and contained in Codex codes of practice unless, in specific cases, the Commission has decided otherwise, or

(iii) to draft provisions on hygiene applicable to specific food items or food groups, whether coming within the terms of reference of a Codex commodity committee or not;

(c) to consider specific hygiene problems assigned to it by the Commission.

CODEX COMMITTEE ON FOOD LABELLING

Host Government: Canada

Sessions:

1st, Ottawa, 21-25 June 1965
2nd, Ottawa, 25-29 July 1966
3rd, Ottawa, 26-30 June 1967
4th, Ottawa, 23-28 September 1968
5th, Rome, 6 April 1970
6th, Geneva, 28-29 June 1971
7th, Ottawa, 5-10 June 1972
8th, Ottawa, 28 May - 1 June 1973
9th, Rome, 26-27 June 1974
10th, Ottawa, 26-30 May 1975
11th, Rome, 25-26 March 1976
12th, Ottawa, 16-20 May 1977

---

[1] The term "hygiene" includes, where necessary, microbiological specifications for food and associated methodology.

13th,	Ottawa, 16-20 July 1979
14th,	Rome, 28-30 November 1979
15th,	Ottawa, 10-14 November 1980
16th,	Ottawa, 17-21 May 1982
17th,	Ottawa, 12-21 October 1983
18th,	Ottawa, 11-18 March 1985
19th,	Ottawa, 9-13 March 1987
20th,	Ottawa, 3-7 April 1989
21st,	Ottawa, 11-15 March 1991
22nd,	Ottawa, 26-30 April 1993
23rd,	Ottawa, 24-28 October 1994
24th,	Ottawa, 14-17 May 1996
25th,	Ottawa, 15-18 April 1997

Terms of reference:

(a) to draft provisions on labelling applicable to all foods;

(b) to consider, amend if necessary, and endorse draft specific provisions on labelling prepared by the Codex Committees drafting standards, codes of practice and guidelines;

(c) to study specific labelling problems assigned to it by the Commission;

(d) to study problems associated with the advertisement of food with particular reference to claims and misleading descriptions.

CODEX COMMITTEE ON METHODS OF ANALYSIS AND SAMPLING

Host Government: Federal Republic of Germany (1st to 6th sessions), Hungary

Sessions:

1st,	Berlin, 23-24 September 1965
2nd,	Berlin, 20-23 September 1966
3rd,	Berlin, 24-27 October 1967
4th,	Berlin, 11-15 November 1968
5th,	Cologne, 1-6 December 1969
6th,	Bonn Bad Godesberg, 24-28 January 1971
7th,	Budapest, 12-18 September 1972
8th,	Budapest, 3-7 September 1973
9th,	Budapest, 27-31 October 1975
10th,	Budapest, 24-28 October 1977

11th,	Budapest, 2-6 July 1979
12th,	Budapest, 11-15 May 1981
13th,	Budapest, 29 November - 3 December 1982
14th,	Budapest, 26-30 November 1984
15th,	Budapest, 10-14 November 1986
16th,	Budapest, 14-19 November 1988
17th,	Budapest, 8-12 April 1991
18th,	Budapest, 9-13 November 1992
19th,	Budapest, 21-25 March 1994
20th,	Budapest, 2-6 October 1995
21st,	Budapest, 10-14 March 1997

Terms of reference:

(a) to define the criteria appropriate to Codex Methods of Analysis and Sampling;

(b) to serve as a coordinating body for Codex with other international groups working in methods of analysis and sampling and quality assurance systems for laboratories;

(c) to specify, on the basis of final recommendations submitted to it by the other bodies referred to in (b) above, Reference Methods of Analysis and Sampling appropriate to Codex Standards which are generally applicable to a number of foods;

(d) to consider, amend, if necessary, and endorse, as appropriate, methods of analysis and sampling proposed by Codex (Commodity) Committees, except that methods of analysis and sampling for residues of pesticides or veterinary drugs in food, the assessment of micro biological quality and safety in food, and the assessment of specifications for food additives, do not fall within the terms of reference of this Committee;

(e) to elaborate sampling plans and procedures, as may be required;

(f) to consider specific sampling and analysis problems submitted to it by the Commission or any of its Committees.

(g) to define procedures, protocols, guidelines or related texts for the assessment of food laboratory proficiency, as well as quality assurance systems for laboratories.

CODEX COMMITTEE ON PESTICIDE RESIDUES

Host Government: Netherlands

Sessions:

1st,	The Hague, 17-21 January 1966
2nd,	The Hague, 18-22 September 1967
3rd,	Arnhem, 30 September-4 October 1968
4th,	Arnhem, 6-14 October 1969
5th,	The Hague, 28 September - 6 October 1970
6th,	The Hague, 16-23 October 1972
7th,	The Hague, 4-9 February 1974
8th,	The Hague, 3-8 March 1975
9th,	The Hague, 14-21 February 1977
10th,	The Hague, 29 May - 5 June 1978
11th,	The Hague, 11-18 June 1979
12th,	The Hague, 2-9 June 1980
13th,	The Hague, 15-20 June 1981
14th,	The Hague, 14-21 June 1982
15th,	The Hague, 3-10 October 1983
16th,	The Hague, 24 May - 4 June 1984
17th,	The Hague, 25 March - 1 April 1985
18th,	The Hague, 21-28 April 1986
19th,	The Hague, 6-13 April 1987
20th,	The Hague, 18-25 April 1988
21st,	The Hague, 10-17 April 1989
22nd,	The Hague, 23-30 April 1990
23rd,	The Hague, 15-22 April 1991
24th,	The Hague, 6-13 April 1992
25th,	Havana, Cuba, 19-26 April 1993
26th,	The Hague, 11-18 April 1994
27th,	The Hague, 24 April-1 May 1995
28th,	The Hague, 15-20 April 1996
29th,	The Hague, 7-12 April 1997

Terms of reference:

(a) to establish maximum limits for pesticide residues in specific food items or in groups of food;

(b) to establish maximum limits for pesticide residues in certain animal feeding stuffs moving in international trade where this is justified for reasons of protection of human health;

(c) to prepare priority lists of pesticides for evaluation by the Joint FAO/WHO Meeting on Pesticide Residues (JMPR);

(d) to consider methods of sampling and analysis for the determination of pesticide residues in food and feed;

(e) to consider other matters in relation to the safety of food and feed containing pesticide residues; and

(f) to establish maximum limits for environmental and industrial contaminants showing chemical or other similarity to pesticides, in specific food items or groups of food.

CODEX COMMITTEE ON RESIDUES OF VETERINARY DRUGS IN FOODS

Host Government: United States of America

Sessions:

1st,	Washington, D.C. 27-31 October, 1986
2nd,	Washington, D.C. 30 November - 4 December 1987
3rd,	Washington, D.C. 31 October - 4 November 1988
4th,	Washington, D.C. 24-27 October 1989
5th,	Washington, D.C. 16-19 October 1990
6th,	Washington, D.C. 22-25 October 1991
7th,	Washington, D.C., 20-23 October 1992
8th,	Washington, D.C., 7-10 June 1994
9th,	Washington, D.C., 5-8 December 1995
10th,	San José, 29 October - 1 November 1996

Terms of reference:

(a) to determine priorities for the consideration of residues of veterinary drugs in foods;

(b) to recommend maximum levels of such substances;

(c) to develop codes of practice as may be required;

(d) to determine criteria for analytical methods used for the control of veterinary drug residues in foods.

CODEX COMMITTEE ON FOOD IMPORT AND EXPORT CERTIFICATION AND INSPECTION SYSTEMS

Host Government, Australia

Sessions:

1st,	Canberra, 21-25 September 1992
2nd,	Canberra, 29 November-3 December 1993
3rd,	Canberra, 27 February-3 March 1995
4th,	Sydney, 19-23 February 1996
5th,	Sydney, 17-21 February 1997

Terms of reference:

(a) to develop principles and guidelines for food import and export inspection and certification systems with a view to harmonising methods and procedures which protect the health of consumers, ensure fair trading practices and facilitate international trade in foodstuffs;

(b) to develop principles and guidelines for the application of measures by the competent authorities of exporting and importing countries to provide assurance where necessary that foodstuffs comply with requirements, especially statutory health requirements;

(c) to develop guidelines for the utilisation, as and when appropriate, of quality assurance systems[1] to ensure that foodstuffs conform with requirements and to promote the recognition of these systems in facilitating trade in food products under bilateral/multilateral arrangements by countries;

(d) to develop guidelines and criteria with respect to format, declarations and language of such official certificates as countries may require with a view towards international harmonization;

(e) to make recommendations for information exchange in relation to food import/export control;

(f) to consult as necessary with other international groups working on matters related to food inspection and certification systems;

---

[1] *Quality assurance* means all those planned and systematic actions necessary to provide adequate confidence that a product or service will satisfy given requirements for quality (ISO-8402 Quality - Vocabulary)

Procedural Manual Page 95

(g) to consider other matters assigned to it by the Commission in relation to food inspection and certification systems.

CODEX COMMITTEE ON NUTRITION AND FOODS FOR SPECIAL DIETARY USES

Host Government: Federal Republic of Germany

Sessions:

1st,	Freiburgh in Breisgau, 2-5 May 1966
2nd,	Freiburgh in Breisgau, 6-10 November 1967
3rd,	Cologne, 14-18 October 1968
4th,	Cologne, 3-7 November 1969
5th,	Bonn, 30 November-4 December 1970
6th,	Bonn, 6-10 December 1971
7th,	Cologne, 10-14 October 1972
8th,	Bonn Bad Godesberg, 9-14 September 1974
9th,	Bonn, 22-26 September 1975
10th,	Bonn, 28 February - 4 March 1977
11th,	Bonn Bad Godesberg, 23-27 October 1978
12th,	Bonn Bad Godesberg, 29 September - 3 October 1980
13th,	Bonn Bad Godesberg, 20-24 September 1982
14th,	Bonn Bad Godesberg, 24 January - 1 February 1985
15th,	Bonn Bad Godesberg, 12-16 January 1987
16th,	Bonn Bad Godesberg, 29 September - 7 October 1988
17th,	Bonn-Bad Godesberg, 18-22 February 1991
18th,	Bonn-Bad Godesberg, 28 September - 2 October 1992
19th,	Bonn-Bad Godesberg, 27-31 March 1995
20th,	Bonn-Bad Godesberg, 7-11 October 1996

Terms of reference:

(a) to study specific nutritional problems assigned to it by the Commission and advise the Commission on general nutrition issues;

(b) to draft general provisions, as appropriate, concerning the nutritional aspects of all foods;

(c) to developed standards, guidelines or related texts for foods for special dietary uses, in cooperation with other committees where necessary;

(d) to consider, amend if necessary, and endorse provisions on nutritional aspects proposed for inclusion Codex standards, guidelines and related texts.

CODEX COMMITTEE ON COCOA PRODUCTS AND CHOCOLATE

Host Government: Switzerland

Sessions:

1st,	Neuchâtel, 5-6 November 1963
2nd,	Montreux, 22-24 April 1964
3rd,	Zürich, 10-12 March 1965
4th,	Berne, 15-17 March 1966
5th,	Lugano, 9-12 May 1967
6th,	Montreux, 2-5 July 1968
7th,	Horgen, (Zürich), 23-27 June 1969
8th,	Lucerne, 29 June - 3 July 1970
9th,	Neuchâtel, 27 September - 1 October 1971
10th,	Lausanne, 7-11 May 1973
11th,	Zürich, 2-6 December 1974
12th,	Bienne, 1-5 November 1976
13th,	Aarau, 2-6 April 1979
14th,	Lausanne, 21-25 April 1980
15th,	Neuchâtel, 29 March - 2 April 1982
16th,	Thun, 30 September - 2 October 1996

Adjourned *sine die*

Terms of reference:

To elaborate world wide standards for cocoa products and chocolate.

CODEX COMMITTEE ON SUGARS

Host Government: United Kingdom

Sessions:

1st,	London, 3-5 March 1964
2nd,	London, 2-4 March 1965
3rd,	London, 1-3 March 1966
4th,	London, 18-21 April 1967
5th,	London, 10-12 September 1968
6th,	London, 19-22 March 1974

Adjourned *sine die*

Terms of reference:

To elaborate world wide standards for all types of sugars and sugar products.

CODEX COMMITTEE ON PROCESSED FRUITS AND VEGETABLES

Host Government: *United States of America*

Sessions:

1st,	Washington, D.C., 29-30 May 1964
2nd,	Rome, 8-11 June 1965
3rd,	Rome, 6-10 June 1966
4th,	Washington, D.C., 19-23 June 1967
5th,	Washington, D.C., 13-17 May 1968
6th,	Washington, D.C., 12-16 May 1969
7th,	Washington, D.C., 1-5 June 1970
8th,	Washington, D.C., 7-11 June 1971
9th,	Washington, D.C., 12-16 June 1972
10th,	Washington, D.C., 21-25 May 1973
11th,	Washington, D.C., 3-7 June 1974
12th,	Washington, D.C., 19-23 May 1975
13th,	Washington, D.C., 9-13 May 1977
14th,	Washington, D.C., 25-29 September 1978
15th,	Washington, D.C., 17-21 March 1980
16th,	Washington, D.C., 22-26 March 1982
17th,	Washington, D.C., 13-17 February 1984
18th,	Washington, D.C., 10-14 March 1986

Adjourned *sine die.*

Terms of reference:

To elaborate world wide standards for all types of processed fruits and vegetables including dried products, canned dried peas and beans, jams and jellies, but not dried prunes, or fruit and vegetable juices.

CODEX COMMITTEE ON FATS AND OILS

Host Government: *United Kingdom*

Sessions:

1st,	London, 25-27 February 1964

2nd,	London, 6-8 April 1965
3rd,	London, 29 March - 1 April 1966
4th,	London, 24-28 April 1967
5th,	London, 16-20 September 1968
6th,	Madrid, 17-20 November 1969
7th,	London, 25-29 March 1974
8th,	London, 24-28 November 1975
9th,	London, 28 November - 2 December 1977
10th,	London, 4-8 December 1978
11th,	London, 23-27 June 1980
12th,	London, 19-23 April 1982
13th,	London, 23-27 February 1987
14th,	London, 27 September - 1 October 1993
15th,	London, 4-8 November 1996

Terms of reference:

To elaborate world wide standards for fats and oils of animal, vegetable and marine origin including margarine and olive oil.

CODEX COMMITTEE ON MEAT

Host Government: Federal Republic of Germany

Sessions:

1st,	Kulmbach, 28-30 October 1965
2nd,	Kulmbach, 5-8 July 1966
3rd,	Kulmbach, 15-17 November 1967
4th,	Kulmbach, 18-20 June 1969
5th,	Bonn, 16-20 November 1970
6th,	Kulmbach, 1-5 November 1971
7th,	Kulmbach, 25-29 June 1973

Dissolved by the 16th Session of the Commission in 1985.

Terms of reference:

To elaborate world wide standards and/or descriptive texts and/or codes of practice as may seem appropriate for the classification, description and grading of carcasses and cuts of beef, veal, mutton, lamb and pork.

CODEX COMMITTEE ON MEAT HYGIENE

Host Government: New Zealand

Sessions:

1st,	London, 10-15 April 1972
2nd,	London, 18-22 June 1973
3rd,	London, 25-29 November 1974
4th,	London, 18-22 May 1981
5th,	London, 11-15 October 1982
6th,	Rome, 14-18 October 1991
7th,	Rome, 29 March - 2 April 1993

Adjourned *sine die.*

Terms of reference:

To elaborate world wide standards and/or codes of practice as may seem appropriate for meat hygiene, excluding poultry meat.

CODEX COMMITTEE ON PROCESSED MEAT AND POULTRY PRODUCTS

Host Government: Denmark

Sessions:

1st,	Kulmbach, 4-5 July 1966
2nd,	Copenhagen, 2-6 October 1967
3rd,	Copenhagen, 24-28 June 1968
4th,	Copenhagen, 9-13 June 1969
5th,	Copenhagen, 23-27 November 1970
6th,	Copenhagen, 17-21 April 1972
7th,	Copenhagen, 3-7 December 1973
8th,	Copenhagen, 10-14 March 1975
9th,	Copenhagen, 29 November - 3 December 1976
10th,	Copenhagen, 20-24 November 1978
11th,	Copenhagen, 22-26 September 1980
12th,	Copenhagen, 4-8 October 1982
13th,	Copenhagen, 23-26 October 1984
14th,	Copenhagen, 12-16 September 1988
15th,	Copenhagen, 8-12 October 1990

Adjourned *sine die.*

Terms of reference:

To elaborate world wide standards for processed meat products, including consumer packaged meat, and for processed poultry meat products.

CODEX COMMITTEE ON FISH AND FISHERY PRODUCTS

Host Government: Norway

Sessions:

1st,	Bergen, 29 August - 2 September 1966
2nd,	Bergen, 9-13 October 1967
3rd,	Bergen, 7-11 October 1968
4th,	Bergen, 29 September 8 - October 1969
5th,	Bergen, 5-10 October 1970
6th,	Bergen, 4-8 October 1971
7th,	Bergen, 2-7 October 1972
8th,	Bergen, 1-6 October 1973
9th,	Bergen, 30 September - 5 October 1974
10th,	Bergen, 29 September - 4 October 1975
11th,	Bergen, 27 September - 2 October 1976
12th,	Bergen, 3-8 October 1977
13th,	Bergen, 7-11 May 1979
14th,	Bergen, 5-10 May 1980
15th,	Bergen, 3-8 May 1982
16th,	Bergen, 7-11 May 1984
17th,	Oslo, 5-9 May 1986
18th,	Bergen, 2-6 May 1988
19th,	Bergen, 11-15 June 1990
20th,	Bergen, 1-5 June 1992
21st,	Bergen, 2-6 May 1994
22nd,	Bergen, 6-10 May 1996

Terms of reference:

To elaborate world wide standards for fresh, frozen (including quick frozen) or otherwise processed fish, crustaceans and molluscs.

CODEX COMMITTEE ON EDIBLE ICES

Host Government: Sweden

Sessions:

1st,	Stockholm, 18-22 February 1974
2nd,	Stockholm, 23-27 June 1975
3rd,	Stockholm, 11-15 October 1976

Abolished by the 22nd Session of the Commission, 1997.

Terms of reference:

To elaborate world wide standards as appropriate for all types of edible ices, including mixes and powders used for their manufacture.

CODEX COMMITTEE ON SOUPS AND BROTHS

Host Government: Switzerland

Sessions:

1st,	Berne, 3-7 November 1975
2nd,	St. Gallen, 7-11 November 1977

Adjourned *sine die.*

Terms of reference:

To elaborate world wide standards for soups, broths, bouillons and consommés.

CODEX COMMITTEE ON CEREALS, PULSES AND LEGUMES

Host Government: United States of America

Sessions:

1st,	Washington, D.C., 24-28 March 1980
2nd,	Washington, D.C., 27 April - 1 May 1981
3rd,	Washington, D.C., 25-29 October 1982
4th,	Washington, D.C., 24-28 September 1984
5th,	Washington, D.C., 17-21 March 1986
6th,	Washington, D.C., 24-28 October 1988
7th,	Washington, D.C., 22-26 October 1990

8th, Washington, D.C., 26-30 October 1992
9th, Washington, D.C., 31 October - 4 November 1994

Adjourned *sine die.*

Terms of reference:

To elaborate world wide standards and/or codes of practice as may be appropriate for cereals, pulses, legumes and their products.

CODEX COMMITTEE ON VEGETABLE PROTEINS

Host Government: Canada

Sessions:

1st, Ottawa, 3-7 November 1980
2nd, Ottawa, 1-5 March 1983
3rd, Ottawa, 6-10 February 1984
4th, Havana, 2-6 February 1987
5th, Ottawa, 6-10 February 1989

Adjourned *sine die.*

Terms of reference:

To elaborate definitions and world wide standards for vegetable protein products deriving from any member of the plant kingdom as they come into use for human consumption, and to elaborate guidelines on utilization of such vegetable protein products in the food supply system, on nutritional requirements and safety, on labelling and on other aspects as may seem appropriate.

CODEX COMMITTEE ON FRESH FRUITS AND VEGETABLES

Established by the 17th Session of the Commission (1987) as the Codex Committee on Tropical Fresh Fruits and Vegetables. Its name and Terms of Reference were amended by the 21st Session of the Commission (1995).

Host Government: Mexico

Sessions:

1st, Mexico City, 6-10 June 1988
2nd, Mexico City, 5-9 March 1990

3rd, Mexico City, 23-27 September 1991
4th, Mexico City, 1-5 February 1993
5th, Mexico City, 5-9 September 1994
6th, Mexico City, 29 January - 2 February 1996

Terms of Reference:

(a) to elaborate world wide standards and codes of practice as may be appropriate for fresh fruits and vegetables;

(b) to consult with the UN/ECE Working Party on Standardization of Perishable Produce in the elaboration of world wide standards and codes of practice with particular regard to ensuring that there is no duplication of standards or codes of practice and that they follow the same broad format[1];

(c) to consult, as necessary, with other international organizations which are active in the area of standardization of fresh fruits and vegetables.

---

[1] The Working Party on Standardization of Perishable Produce of the United Nations Economic Commission for Europe:

1. may recommend that a world wide Codex standard for fresh fruits and vegetables should be elaborated and submit its recommendation either to the Codex Committee on Fresh Fruits and Vegetables for consideration or to the Commission for approval;

2. may prepare "proposed draft standards" for fresh fruits or vegetables at the request of the Codex Committee on Fresh Fruits and Vegetables or of the Commission for distribution by the Codex Secretariat at Step 3 of the Codex Procedure, and for further action by the Codex Committee on Fresh Fruits and Vegetables;

3. may wish to consider "proposed draft standards" and "draft standards" for fresh fruits and vegetables and transmit comments on them to the Codex Committee on Fresh Fruits and Vegetables at Steps 3 and 6 of the Codex Procedure; and

4. may perform specific tasks in relation to the elaboration of standards for fresh fruits and vegetables at the request of the Codex Committee on Fresh Fruits and Vegetables.

Codex "proposed draft standards" and "draft standards" for fresh fruits and vegetables at Steps 3 and 6 of the Codex Procedure should be submitted to the UN/ECE Secretariat for obtaining comments.

CODEX COMMITTEE ON MILK AND MILK PRODUCTS

Host Government: New Zealand

Sessions:

1st, Rome, 28 November - 2 December 1994
2nd, Rome, 27-31 May 1996

Terms of reference:

To elaborate international codes and standards for milk and milk products within the framework of the Codex Alimentarius and the Code of Principles concerning Milk and Milk Products.

CODEX COMMITTEE ON NATURAL MINERAL WATERS

Host Government: Switzerland

Sessions:

1st, Badan/Aarzan, 24-25 February 1966
2nd, Montreux, 6-7 July 1967
3rd, Bad Ragaz, - 9 May 1968
4th, Vienna, 12-13 June 1972
5th, Thun, 3-5 October 1996

Adjourned *sine die.*

Terms of reference:

To elaborate regional standards for natural mineral waters.

Note: The Committee was established by the Commission as a Regional (European) Codex Committee, but has since been allocated the task of elaborating world-wide standards for natural mineral waters.

SUBSIDIARY BODIES UNDER RULE IX.1(B)(II)

FAO/WHO COORDINATING COMMITTEE FOR AFRICA

Membership:

Membership of the Committee is open to all Member Nations and Associate Members of FAO and/or WHO which are members of the Codex Alimentarius Commission, within the geographic location of Africa.

Terms of reference:

(a) defines the problems and needs of the region concerning food standards and food control;

(b) promotes within the Committee contacts for the mutual exchange of information on proposed regulatory initiatives and problems arising from food control and stimulates the strengthening of food control infrastructures;

(c) recommends to the Commission the development of world wide standards for products of interest to the region, including products considered by the Committee to have an international market potential in the future;

(d) develops regional standards for food products moving exclusively or almost exclusively in intra regional trade;

(e) draws the attention of the Commission to any aspects of the Commission's work of particular significance to the region;

(f) promotes coordination of all regional food standards work undertaken by international governmental and non-governmental organizations within the region;

(g) exercises a general coordinating role for the region and such other functions as may be entrusted to it by the Commission;

(h) promotes the acceptance of Codex standards and maximum limits for residues by member countries.

Sessions:

1st,	Rome, 24-27 June 1974
2nd,	Accra, 15-19 September 1975
3rd,	Accra, 26-30 September 1977
4th,	Dakar, 3-7 September 1979
5th,	Dakar, 25-29 May 1981
6th,	Nairobi, 31 October - 5 November 1983
7th,	Nairobi, 12-18 February 1985
8th,	Cairo, 29 November - 3 December 1988
9th,	Cairo, 3-7 December 1990
10th,	Abuja, 3-6 November 1992
11th,	Abuja, 8-11 May 1995
12th,	Harare, 19-22 November 1996

FAO/WHO COORDINATING COMMITTEE FOR ASIA

Membership:

Membership of the Committee is open to all Member Nations and Associate Members of FAO and/or WHO which are members of the Codex Alimentarius Commission, within the geographic location of Asia.

Terms of reference:

(a) defines the problems and needs of the region concerning food standards and food control;

(b) promotes within the Committee contacts for the mutual exchange of information on proposed regulatory initiatives and problems arising from food control and stimulates the strengthening of food control infrastructures;

(c) recommends to the Commission the development of world wide standards for products of interest to the region, including products considered by the Committee to have an international market potential in the future;

(d) develops regional standards for food products moving exclusively or almost exclusively in intra regional trade;

(e) draws the attention of the Commission to any aspects of the Commission's work of particular significance to the region;

(f) promotes coordination of all regional food standards work undertaken by international governmental and non-governmental organizations within the region;

(g) exercises a general coordinating role for the region and such other functions as may be entrusted to it by the Commission;

(h) promotes the acceptance of Codex standards and maximum limits for residues by member countries.

Sessions:

    1st,   New Delhi, 10-16 January 1977
    2nd,   Manila, 20-26 March 1979
    3rd,   Colombo, 2-8 February 1982
    4th,   Phetchburi, 28 February - 5 March 1984
    5th,   Yogyakarta, 8-14 April 1986
    6th,   Denpasar, 26 January - 1 February 1988
    7th,   Chiang-Mai, 5-12 February 1990
    8th,   Kuala Lumpur, 27-31 January 1992
    9th,   Beijing, 24-27 May 1994

10th, Tokyo, 5-8 March 1996

FAO/WHO COORDINATING COMMITTEE FOR EUROPE

Membership:

This Committee is open to all Member Governments of FAO and/or WHO within the geographic area of Europe, including Israel, Turkey and the Russian Federation and its Chairperson is, ex officio, the Coordinator for Europe.

Terms of reference:

(a) defines the problems and needs of the region concerning food standards and food control;

(b) promotes within the Committee contacts for the mutual exchange of information on proposed regulatory initiatives and problems arising from food control and stimulates the strengthening of food control infrastructures;

(c) recommends to the Commission the development of world wide standards for products of interest to the region, including products considered by the Committee to have an international market potential in the future;

(d) develops regional standards for food products moving exclusively or almost exclusively in intra regional trade;

(e) draws the attention of the Commission to any aspects of the Commission's work of particular significance to the region;

(f) promotes coordination of all regional food standards work undertaken by international governmental and non-governmental organizations within the region;

(g) exercises a general coordinating role for the region and such other functions as may be entrusted to it by the Commission, and

(h) promotes the acceptance of Codex standards and maximum limits for residues by member countries.

Sessions:

1st, Berne, 1-2 July 1965
2nd, Rome, 20 October 1965
3rd, Vienna, 24-27 May 1966
4th, Rome, 8 November 1966
5th, Vienna, 6-8 September 1967
6th, Vienna, 4-8 November 1968

7th,	Vienna, 7-10 October 1969
8th,	Vienna, 27-29 October 1971
9th,	Vienna, 14-16 June 1972
10th,	Vienna, 13-17 June 1977
11th,	Innsbruck, 28 May - 1 June 1979
12th,	Innsbruck, 16-20 March 1981
13th,	Innsbruck, 27 September - 1 October 1982
14th,	Thun, 4-8 June 1984
15th,	Thun, 16-20 June 1986
16th,	Vienna, 27 June - 1 July 1988
17th,	Vienna, 28 May - 1 June 1990
18th,	Stockholm, 11-15 May 1992
19th,	Stockholm, 16-20 May 1994
20th,	Uppsala, 23-26 April 1996

FAO/WHO COORDINATING COMMITTEE FOR LATIN AMERICA AND THE CARIBBEAN

Membership:

Membership of the Committee is open to all Member Nations and Associate Members of FAO and/or WHO which are members of the Codex Alimentarius Commission, within the geographic location of Latin America and the Caribbean.

Terms of reference:

(a) defines the problems and needs of the region concerning food standards and food control;

(b) promotes within the Committee contacts for the mutual exchange of information on proposed regulatory initiatives and problems arising from food control and stimulates the strengthening of food control infrastructures;

(c) recommends to the Commission the development of world wide standards for products of interest to the region, including products considered by the Committee to have an international market potential in the future;

(d) develops regional standards for food products moving exclusively or almost exclusively in intra regional trade;

(e) draws the attention of the Commission to any aspects of the Commission's work of particular significance to the region;

(f) promotes coordination of all regional food standards work undertaken by international governmental and non-governmental organizations within the region;

(g) exercises a general coordinating role for the region and such other functions as may be entrusted to it by the Commission, and

(h) promotes the acceptance of Codex standards and maximum limits for residues by member countries.

Sessions:

1st,	Rome, 25-26 March 1976
2nd,	Montevideo, 9-15 December 1980
3rd,	Havana, 27 March - 2 April 1984
4th,	Havana, 17-22 April 1985
5th,	Havana, 11-16 February 1987
6th,	San José, 20-24 February 1989
7th,	San José, 1-10 July 1991
8th,	Brasília, 16-20 March 1993
9th,	Brasília, 3-7 April 1995
10th,	Montevideo, 25-28 February 1997

FAO/WHO COORDINATING COMMITTEE FOR NORTH AMERICA AND THE SOUTH WEST PACIFIC

Membership:

Membership of the Committee is open to all Member Nations and Associate Members of FAO and/or WHO which are members of the Codex Alimentarius Commission, with the geographic locations of North America and the South West Pacific.

Terms of reference:

(a) defines the problems and needs of regions concerning food standards and food control;

(b) promotes within the Committee contacts for the mutual exchange of information on proposed regulatory initiatives and problems arising from food control and stimulates the strengthening of food control infrastructures;

(c) recommends to the Commission the development of world wide standards for products of interest to the regions, including products considered by the Committee to have an international market potential in the future;

(d) develops regional standards for food products moving exclusively or almost exclusively in intra regional trade;

(e) draws the attention of the Commission to any aspects of the Commission's work of particular significance to the regions;

(f) promotes coordination of all regional food standards work undertaken by international governmental and non-governmental organizations within the regions;

(g) exercises a general coordinating role for the regions and such other functions as may be entrusted to it by the Commission, and

(h) promotes the acceptance of Codex standards and maximum limits for residues by member countries.

Sessions:

1st,	Honolulu, 30 April - 4 May 1990
2nd,	Canberra, 2-6 December 1991
3rd,	Vancouver, 31 May - 3 June 1994
4th,	Rotorua, 30 April - 3 May 1996

JOINT ECE/CODEX ALIMENTARIUS GROUPS OF EXPERTS ON STANDARDIZATION[1]

QUICK FROZEN FOODS

Sessions:

1st,	Geneva, 6-10 September 1965
2nd,	Geneva, 5-9 September 1966
3rd,	Rome, 18-22 September 1967
4th,	Geneva, 2-6 September 1968
5th,	Rome, 22-26 September 1969
6th,	Rome, 27-31 July 1970
7th,	Geneva, 6-10 December 1971
8th,	Geneva, 30 April - 4 May 1973
9th,	Rome, 7-11 October 1974

---

[1] These Joint ECE/Codex Alimentarius committees are not subsidiary bodies under any specific rule of the Codex Alimentarius Commission but follow the same procedure as Codex Commodity Committees for the elaboration of Codex standards.

10th,	Geneva, 6-10 October 1975
11th,	Geneva, 14-18 March 1977
12th,	Rome, 30 October - 6 November 1978
13th,	Rome, 15-19 September 1980

Adjourned *sine die*.

Terms of reference:

The Joint ECE/Codex Alimentarius Group of Experts on the Standardization of Quick Frozen Foods will be responsible for the development of standards for quick frozen foods in accordance with the General Principles of the Codex Alimentarius. The Joint Group will be responsible for general considerations, definitions, a framework of individual standards for quick frozen food products and for the actual elaboration of standards for quick frozen food products not specifically allotted by the Commission to another Codex Committee, such as Fish and Fishery Products, Meat, Processed Meat and Poultry Products. Standards drawn up by Codex commodity committees for quick frozen foods should be in accordance with the general standard laid down by the Joint ECE/Codex Alimentarius Group of Experts on the Standardization of Quick Frozen Foods and should, at an appropriate stage, be referred to it for coordination purposes.

FRUIT JUICES

Sessions:

1st,	Geneva, 6-10 April 1964
2nd,	Geneva, 29 March - 2 April 1965
3rd,	Geneva, 21-25 February 1966
4th,	Geneva, 10-14 April 1967
5th,	Rome, 25-29 March 1968
6th,	Geneva, 27-31 October 1969
7th,	Rome, 20-24 July 1970
8th,	Geneva, 8-12 March 1971
9th,	Rome, 20-24 March 1972
10th,	Geneva, 16-20 July 1973
11th,	Rome, 14-18 October 1974
12th,	Geneva, 19-23 July 1976
13th,	Geneva, 26-30 June 1978
14th,	Geneva, 9-13 June 1980
15th,	Rome, 8-12 February 1982
16th,	Geneva, 30 April - 4 May 1984

17th, Rome, 26-30 May 1986
18th, Geneva, 16-20 May 1988
19th, Rome 12-16 November 1990

Adjourned *sine die.*

Terms of reference:

To elaborate world wide standards for fruit juices, concentrated fruit juices and nectars.

MEMBERS OF THE CODEX ALIMENTARIUS COMMISSION

Africa
1. Algeria
2. Angola
3. Benin
4. Botswana
5. Burkina Faso
6. Burundi
7. Cameroon
8. Cape Verde
9. Central African Republic
10. Chad
11. Congo, Democratic Republic of
12. Congo, Republic of
13. Côte d'Ivoire
14. Egypt
15. Equatorial Guinea
16. Eritrea
17. Ethiopia
18. Gabon
19. Gambia
20. Ghana
21. Guinea
22. Guinea Bissau
23. Kenya
24. Lesotho
25. Liberia
26. Libyan Arab Jamahiriya
27. Madagascar
28. Malawi
29. Mauritania
30. Mauritius
31. Morocco
32. Mozambique
33. Niger
34. Nigeria
35. Rwanda
36. Senegal
37. Seychelles
38. Sierra Leone
39. South Africa
40. Sudan
41. Swaziland
42. Togo
43. Tunisia
44. Uganda
45. United Republic of Tanzania
46. Zambia
47. Zimbabwe

Asia
48. Bahrain
49. Bangladesh
50. Brunei Darussalam
51. Cambodia
52. China
53. Democratic People's Republic of Korea
54. India
55. Indonesia
56. Iran (Islamic Republic of)
57. Iraq
58. Japan

59. Jordan
60. Kuwait
61. Laos
62. Lebanon
63. Malaysia
64. Mongolia
65. Myanmar
66. Nepal
67. Oman
68. Pakistan
69. Philippines
70. Qatar
71. Republic of Korea
72. Saudi Arabia
73. Singapore
74. Sri Lanka
75. Syrian Arab Republic
76. Thailand
77. Turkey
78. United Arab Emirates
79. Viet Nam
80. Yemen

Europe
81. Albania
82. Armenia
83. Austria
84. Belgium
85. Bulgaria
86. Croatia
87. Cyprus
88. Czech Republic
89. Denmark
90. Estonia
91. Finland
92. France
93. Germany
94. Greece
95. Hungary
96. Iceland
97. Ireland
98. Israel
99. Italy
100. Latvia
101. Lithuania
102. Luxembourg
103. Malta
104. Netherlands
105. Norway
106. Poland
107. Portugal
108. Romania
109. Russian Federation
110. Slovak Republic
111. Slovenia
112. Spain
113. Sweden
114. Switzerland
115. The Former Yugoslav Republic of Macedonia
116. United Kingdom
117. Yugoslavia

Latin America and the Caribbean
118. Antigua and Barbuda
119. Argentina
120. Barbados
121. Belize
122. Bolivia
123. Brazil

124. Chile
125. Colombia
126. Costa Rica
127. Cuba
128. Dominica
129. Dominican Republic
130. Ecuador
131. El Salvador
132. Grenada
133. Guatemala
134. Guyana
135. Haiti
136. Honduras
137. Jamaica
138. Mexico
139. Nicaragua
140. Panama
141. Paraguay
142. Peru
143. Saint Kitts and Nevis
144. Saint Lucia
145. Suriname
146. Trinidad and Tobago
147. Uruguay
148. Venezuela

North America

149. Canada
150. United States of America

South-West Pacific

151. Australia
152. Fiji
153. Kiribati
154. Micronesia, Federated States of
155. New Zealand
156. Papua New Guinea
157. Samoa
158. Vanuatu

LIST OF CODEX CONTACT POINTS[1]

ALBANIA	Directorate of Food Quality and Inspection Documentation Centre Ministry of Agriculture and Food Skenderbe sq. Tirana Fax: +355 42 279244
ALGERIA	Ministère de l'Economie Direction générale de la concurrence et des prix Direction de la qualité et de la consommation Palais du Gouvernement Alger
ANGOLA	Mr. Estevâo Miguel de Carvalho Director, Gabinete Tecnico Ministerio da Agricultura e Desenvolvimento Rural C.P. 527 Luanda Fax: +244 2320553/ 321943
ANTIGUA AND BARBUDA	Director, Antigua and Barbuda Bureau of Standards (ABBS) P.O. Box 1550 Redcliffe Street St. John's Antigua Tel: +854 462 1625 Fax: +854 462 1532

---

[1] This list is subject to change. Contact Points of new Members are notified by Circular Letter. Revised lists are circulated at regular intervals and an up-dated list is maintained on the World-Wide Web at the following address: http://www.fao.org/waicent/faoinfo/economic/esn/codex/codex.htm.

ARGENTINA	Secretaría de Agricultura, Pesca y Alimentación, Subsecretaría de Alimentación Paseo Colón 982, 1° Piso, Oficina 68 (1063) Buenos Aires
	Tel: +54 1 349 2044/ 349 2186/ 349 2165 Fax: +54 1 349 2097/ 349 2162 Email: CODEX@sagyp.mecon.ar
ARMENIA	Dr. A. Malkhassian Director, State Enterprise "Paren" Ministry of Food and Purchase Gorvetca str. 4 375023 Yerevan
	Tel: +374 2 52 46 86 52 46 87 Fax: +374 2 52 88 43 Telex: 243338 VOLT SU
AUSTRALIA	Mr. Digby Gascoine Director, Policy and International Division Australian Quarantine and Inspection Service GPO Box 858 Canberra ACT 2601
	Tel: +61 6272 5584 Fax: +61 6272 3103 Email: codex.contact@dpie.gov.au
AUSTRIA	Bundesministerium für Land und Forstwirtschaft (Div. III/A/3) Stubenring 12 A-1011 Vienna
	Tel: +43 222 71100 Fax: +43 222 7135413/ 71100 2892
BAHRAIN	Dr. Rifa'at Abdul Hameed Director of Public Health P.O. Box 42 Manama
	Fax: +973 25 25 69

BANGLADESH	Director-General Bangladesh Standards and Testing Institution (BSTI) 116/A, Tejgaon Industrial Area Dhaka 8
BARBADOS	Director Barbados National Standards Institution "Flodden", Culloden Road St. Michael
BELGIUM	Comité belge du Codex Alimentarius Service du commerce international des matières premières et produits agricoles (B14) Ministère des relations extérieures rue Quatre Bras, 2 B-1000 Bruxelles Tel: +32 2 516 8299 516 8294 Fax: +32 2 516 88 27 Telex: 23979
BELIZE	The Director Bureau of Standards 53 Regent Street P.O. Box No. 1647 Belize City Tel: +501 2 72 314/ 2 71 584 Fax: +501 2 74 984
BENIN	Secretariat de la Commission nationale du Codex Alimentarius Direction de l'Alimentation et de la Nutrition appliquée (DANA) Ministère du Développement Rural B.P. No. 295, Porto Novo Tel: +229 21 26 70

BOLIVIA	Lic. Fernando Fernández Director Ejecutivo, Instituto Boliviano de Normalización y Calidad (IBNORCA), Av. Camacho 1488 Esquina Bueno - Casilla No. 5034 La Paz Tel: +591 2 317262/ 319185 Fax: +591 2 317262
BOTSWANA	The Head, National Food Control Laboratory Ministry of Health Private Bag 00269 Gaborone Fax: +267 374 354
BRAZIL	DIE - Divisao de Organismos Internacionales Especializados Ministerio das Relacoes Exteriores Expl. dos Ministerios, Pal. do Itamaraty, Anexo I - Sala 418 70.170 Brasilia Tel: +55 61 211 6328/ 6329/6330 Fax: +55 61 322 0860 Telex: 1319 MNRE BR
BRUNEI DARUSSALAM	Dr. Mohamed Yussof Bin Haji Mohiddin Department of Agriculture Bandar Seri Begawan Brunei 2059 Fax: +673 2 38 2226/38 1639
BULGARIA	Monsieur le Chef de la Section de la Commission du Codex Alimentarius Union nationale agro-industrielle 55, boul. Hristo Botev 1000 Sofia Tel: +359 2 8531/ Ext.617 Fax: +359 2 800 655
BURKINA FASO	Ministre du développement rural Ministère du développement rural P.O. Box 7010 Ouagadougou

BURUNDI   Bureau Burundais de normalisation et
                    contrôle de la qualité "BNN"
                    B.P. 3535, Bujumbura

                    Tel: +257 2 22 1815 22 1577

CAMBODIA   Mr. Lim Thearith
                    Assistant Quality Control Service
                    KAMCONTROL, 50E/144 Street
                    Phnom-penh

                    Tel: +855 2 3485
                    Fax: +855 2 3426166

CAMEROON   Ministère du Développement Industriel et
                    Commercial
                    Codex Alimentarius - Service Central de
                    Liaison
                    (Attention: Mme. M.M. Nguidjoi)
                    Yaoundé

                    Tel: +237 22 09 16
                    Fax: +237 22 27 04
                    Telex: 8638 KN

CANADA   Mr. Ron B, Burke, Deputy Director
                    Bureau of Food Regulatory, Inter-national and
                    Interagency Affairs
                    Food Directorate, Health Protection
                    Branch, Health Canada
                    Room 200, H.P.B. Building
                    Tunney's Pasture
                    Ottawa, Ontario K1A OL2

                    Tel: +1 613 957 1748
                    Fax: +1 613 941 3537 952 7767
                    Email: Santina_Scalzo@isdtcp3.hwc.ca

CAPE VERDE   Gabinete de Estudos e Planeamento
                    Ministerio de Pescas, de Agricultura
                    y Animation Rurale
                    Caixa Postal 115
                    Cidade de Praia

                    Fax: +238 64054

CENTRAL AFRICAN REPUBLIC	Ministre des Eaux, des Fôrets, de la Chasse, de la Pêche, chargé de l'Environnement, Ministère des Eaux, des Fôrets, de la Chasse, de la Pêche, chargé de l'Environnement Bangui
CHAD	Direction du génie sanitaire et de l'assainissement Sous-direction de l'assainissement B.P. 440 N'Djamena
	Fax: +235 51 51 85
CHILE	Ministerio de Salud Pública Monjitas 689, 5° Piso Santiago
CHINA	Mr. Xu Guanghua Department of Science and Technology Ministry of Agriculture Beijing
	Fax: +86 10 500 2448
COLOMBIA	Jefe del Programa de Alimentos Subdirección de Ambiente y Salud Ministerio de Salud Carrera 13, No. 32-76, Edificio Urano, Piso 14 Santafé de Bogotá D.C.
	Tel: +57 1 3365066 Ext.1409 Fax: +57 1 3360182
CONGO, DEMOCRATIC REPUBLIC OF	1ère Direction des études et de la politique agricoles Ministère de l'agriculture et du développement rural B.P. 8722 Kinshasa 1
	Tel: +243 12 31126 Telex: 21382 DR KIN ZR

CONGO, REPUBLIC OF	Représentant de la FAO au Congo et à Sao Tomé-et-Principe B.P. 972 Brazzaville Tel: +242 830346 830997 Fax: +242 835502 833987 Telex: FOODAGRI 5348 KG (CONGO)
COSTA RICA	Comité Nacional del Codex Alimentarius Oficina Nacional de Normas y Unidades de Medida Ministerio de Economia, Industria y Comercio A.P. 1736 2050 San José Tel: +506 283 5133 Fax: +506 222 2305/ 283 5133 Telex: 2414 MEC
COTE D'IVOIRE	M. le Secrétaire général Comité national pour l'alimentation et le développement B.P. V 190 Abidjan Tel: +21 49 34
CROATIA	Mrs. Nada Marcovčić Department Chief of Standardization State Office for Standardization and Metrology Ul Grada Vukovara 78 41000 Zagreb Tel: +385 41 63 34 44 Fax: +385 41 53 66 88
CUBA	Sr. Director, Dirección de Relaciones Internacionales Oficina Nacional de Normalización Calle E No. 261 entre 11 y 13 Vedado - La Habana 10400 Tel: +53-7 300022/ 300835/ 300825 Fax: +53-7 338048 Telex: 512245

CYPRUS	Dr. Ioannis G. Karis Director, Cyprus Organization for Standards and Control of Quality Ministry of Commerce and Industry Nicosia Tel: +357 2 30 3441 48 Fax: +357 2 37 51 20
CZECH REPUBLIC	Mr. Pavel Dobrovský Officer-in-Charge odbor 320, Ministerstvo zemědělství ČR Těšnov 17 117 05 Praha 1 Tel: +42 2 2862 869 Fax: +42 2 231 4117
DEMOCRATIC PEOPLE'S REPUBLIC OF KOREA	Director Foodstuffs Institute P.O. Box 901 Pyongyang
DENMARK	Danish Codex Alimentarius Committee Danish Veterinary Service Rolighedsvej 25 DK-1958 Frederiksberg Tel: +45 31 35 81 00 Fax: +45 3536 06 07/ 3536 19 12 Telex: 22 473 vetdir dk
DOMINICAN REPUBLIC	Secretaría de Estado de Salud Pública y Asistencia Social (Sección de Control de Alimentos) Ensanche La Fe Santo Domingo
ECUADOR	Sr. Director General Instituto Ecuatoriano de Normalización Calle Baquerizo Moreno 454 y Almagro (Casilla 17-01-3999) Quito Tel: +593 2 501885 - 501891 Fax: +593 2 567815 Telex: 22687 INEN ED

EGYPT	The President Egyptian Organization for Standardization (EOS) 2 Latin America Street Garden City Cairo Tel: +20 2 354 9720 Fax: +20 2 354 8817 Telex: 932 96 eas un
EL SALVADOR	CONACYT (Consejo Nacional de Ciencia y Tecnología) Urbanización Isidro Menendez Passaje San Antonio No. 51 San Salvador Tel: +503 226 2800 CONACYT Fax: +503 225 6255 CONACYT
EQUATORIAL GUINEA	Mr. Alejandro Ndjoli Mediko Jefe Nacional de Estadísticas Agropecuarias Ministerio de Agricultura, Ganadería, Pesca y Forestal Malabo (Bioko Norte) Fax: +240 9 3178
ERITREA	Dr. Akberom Tedla Head, Eritrean Standards Institution P.O. Box 245 Asmara Tel: +291 1 115624/ 120245 Fax: +291 1 120586
ESTONIA	Ministry of Agriculture Veterinary and Food Department 39/41 Lai str. EE 0100 Tallinn Tel: +372 6.256 210 Fax: +372 6 256 212
ETHIOPIA	Ethiopian Standards Institution P.O. Box 2310 Addis Ababa

FIJI	The Permanent Secretary Ministry of Primary Industries P.O. Box 358 Suva FJ 2290 FIJI FISH FJ
FINLAND	Ministry of Trade and Industry Advisory Committee on Foodstuffs General Secretary Box 230 00171 Helsinki Tel: +358 0 1601 Fax: +358 0 1603666 Telex: 124645 Minco SF
FRANCE	SGCI (Comité interministériel pour le Questions de Coopération Economique Européenne) (Att: Mme Michelle GUNZLE et M. Jean-Luc ANGOT) Carré Austerlitz 2, Boulevard Diderot F-75572 Paris CEDEX 12 Tel: +33 1 44 87 16 00 Fax: +33 1 44 87 16 04 Email: JEAN_LUC.ANGOT@sgci.finances.gouv.fr
GABON	Commission nationale Gabonaise de la FAO Ministère de l'agriculture, de l'élevage et du développement rural B.P. 551 Libreville Tel: +241 763835 Fax: +241 728 275
GAMBIA	The Director of Agriculture Department of Agriculture Ministry of Agriculture Central Bank Building Buckle Street Banjul Fax: +220 228998/ 227994

GERMANY	Mr. Werner Siebenpfeiffer Ministerialdirigent Bundesministerium für Gesundheit Am Propsthof 78a D-53108 Bonn Tel: +49 228 941 4000 Fax: +49 228 941 4940/ 941 4947 Telex: 8 869 355 bmgd
GHANA	The Director Ghana Standards Board P.O. Box M-245 Accra Tel: +233 21 500 065 Fax: +233 21 500 092
GREECE	Direction of Processing Standardization and Quality Control of Agricultural Products of Vegetable Origin Ministry of Agriculture 2 Acharnon St. 104 32 Athens Tel: +30 1 5246364 Fax: +30 1 5240955 Telex: 221701 YGDP GR
GRENADA	Director, Grenada Bureau of Standards Tyrrel Street St. George's Tel: +854 440 5886 Fax: +854 440 4115
GUATEMALA	Dr. L.R. Sandoval Cambára Director-Técnico, Inspección Sanitaria y Control de Alimentos de Origen Animal, DIGESEPE Ministerio de Agricultura, Ganadería y Alimentación Palacio Nacional, Cuidad de Guatemala Fax: +502 2 536 807

GUINEA	M. le Directeur Institut de Normalisation et de Métrologie c/o Ministère de l'Industrie, du Commerce et de l'Artisanat B.P. 468 Conakry
GUINEA BISSAU	Ministère du développement rural et de l'agriculture B.P. 71 1011 Bissau Codex Fax: +245 212 617/ 221 019
GUYANA	Dr. Chatterpaul Ramcharran Director, Guyana National Bureau of Standards 77 W1/2 Hadfield Street Werk-en-Rust P.O. Box 10926 Georgetown Tel: +592 2 57455/ 59013/ 56226 Cable: GUYSTAN
HAITI	M. Raymond Tardieu Direction - normalisation et contrôle de la qualité Ministère du commerce 8, rue Légitime, Champ de Mars Port-au-Prince
HONDURAS	Dra. Georgina Nazar División Control de Alimentos Edificio CESCO-ALIMENTOS Barrio Morazan Tegucigalpa, M.D.C. Tel: +504 32 11 39 Fax: +504 31 27 13

HUNGARY	Dr. Mária Váradi
Scientific Deputy Director
Central Food Research Institute
P.O. Box 393 (Herrman Ottó út 15)
H-1536 Budapest 9

Tel: +36 1 155 8244
Fax: +36 1 155 8991

ICELAND	Mr. J. Gislason
Chief of Division
Environmental and Food Agency
Office of Food and Hygiene
P.O. Box 8080
108 Reykjavik

Tel: +354 1 688848
Fax: +354 1 6818962225 Extern IS

INDIA	Mrs. Debi Mukherjee
Assistant Director General (PFA) cum
Secretary, Central Committee for Food
Standards and Liaison Officer, National
Codex Committee
Directorate General of Health Services
Nirman Bhavan
New Delhi 110 011

Tel: +91 11 3012290
Telex: 31 66119 DGHS IN

INDONESIA	Attn. M. Bambang H. Hadiwiardjo
Dewan Standardisasi Nasional - DSN
(Standardization Council of Indonesia)
Gedung PDIN -LIPI
Jl. Jend. Gatot Subroto 10
(P.O. Box 3123)
Jakarta 12710

Tel: +62 21 522 16 86
Fax: +62 21 520 65 74
Telex: 6 28 75 pdii ia

IRAQ	Central Organization for Standardization and Quality Control Ministry of Planning P.O. Box 13032 Baghdad - Jadria Tel: +964 1 7765180/ 7765181/ 7765182 Fax: +964 1 7765781 Telex: 213505 COSQC IK
IRELAND	The Secretary Irish National FAO Committee Dept. of Agriculture and Fisheries Agriculture House Dublin 2 Tel: +353 1 607 2000 Fax: +353 1 661 62 63 Telex: 93607 AGRI EI
ISLAMIC REPUBLIC OF IRAN	Institute of Standards and Industrial Research of Iran Ministry of Industries P.O. Box 15875-4618 Tehran Fax: +98 21 8802276/261 25015 Telex: 215442 STAN IN
ISRAEL	Mr. Dan Halpern Israel Codex Alimentarius Committee Ministry of Industry and Trade P.O. Box 299 91002 Jerusalem Tel: +972 3 5606146 Fax: +972 3 5605146
ITALY	Sig. Presidente, Comitato Nazionale Italiano per il Codex Alimentarius Direzione Generale della Tutela Economica dei Prodotti Agricoli Via Sallustiana, 10 00187 Roma Tel: +39 6 4665 6510 Fax: +39 6 488 1252/ 474 3971

JAMAICA	Bureau of Standards 6 Winchester Road P.O. Box 113 Kingston 10 Tel: +854 926 3140-6/ 968 2063-71 Fax: +854 929 4736 Telex: 2291 STANBUR JA
JAPAN	Mr. Satoru Iitaka Director, Resources Office, Policy Division, Science and Technology B Science and Technology Agency 2-2-1 Kasumigaseki, Chiyoda-ku 100 Tokyo Tel: +81 3 3581 5271 Fax: +81 3 3581 3079
JORDAN	National Committee for Codex Alimentarius Directorate of Standards Ministry of Industry and Trade P.O. Box 2019 Amman
KENYA	The Director Kenya Bureau of Standards P.O. Box 54974 NHC House, Harambee Avenue Nairobi Tel: +254 2 502210-19 Fax: +254 2 503293 Telex: 25252 'VIWANGO'
KUWAIT	Under Secretary/Assistant for Standards and Metrology Affairs Standards and Metrology Dept. Ministry of Commerce and Industry P.O. Box 2944 - Safat 13030 Kuwait Tel: +965 2465 103/ 2465 101 Fax: +965 2436 638/ 2451 141

LAOS	Dr. Vilayvang Phimmasone Deputy Director Food and Drug Department Ministry of Health Simeuang Road Vientiane Tel: +856-21 214014 Fax: +856-21 214015
LATVIA	Eugenia Pushmucáne Head, State Inspection of Produce Quality Ministry of Agriculture Republiks lavk. 2 Riga 226168, PDP Fax: +371 2 322252
LEBANON	LIBNOR Lebanese Standards Institution P.O. Box 55120 Beirut Tel: +961 1 485927/8 Fax: +961 1 485929
LESOTHO	Mrs. M.N. Mpeta Director, Food and Nutrition Coordinating Office Private Bag A78 Maseru 100
LIBERIA	Mr. Joseph M. Coleman Director of Standards Ministry of Commerce & Industry P.O. Box 10-9041 1000 Monrovia
LIBYAN ARAB JAMAHIRIYA	Director Office of International Cooperation, Secretariat for Agricultural Reclamation and Land Development c/o UNDP Office - P.O. Box 358 Tripoli Fax: +218 21 603449

LITHUANIA	Director, Lithuanian National Nutrition Center Kalvarijų 153 2042 Vilnius Tel: +370 2 778919 Fax: +370 2 778713
LUXEMBOURG	M. François Arendt Ingénieur-chef de Division Laboratoire national de santé 1 A rue Auguste Lumière Luxembourg
MADAGASCAR	Direction de la Qualité et de la Métrologie Légale Ministère du Commerce B.P. 1316 Antananarivo - 101
'MALAWI	The Director Malawi Bureau of Standards P.O. Box 946 - Moirs Road Blantyre Tel: +265 670488 Fax: +265 670756
MALAYSIA	Malaysian National Codex Committee Food Quality and Control Division Ministry of Health, Malaysia 4th Floor, Block E Jalan Dungun, Bukit Damansara 50450 Kuala Lumpur Tel: +60 3 2540088 Fax: +60 3 2537804
MALTA	Mr. J. Samut, Industrial Chemist Standards Laboratory Department of Industry Evans Building, Merchants Street Valletta Tel: +356 221335/ 221873 Fax: +356 236237

MAURITANIA	Centre National d'Hygiène B.P. 695 Nouakchott Tel: +222 253 134 253 175 Fax: +222 253 134
MAURITIUS	The Chief Agricultural Officer Agricultural Services Ministry of Agriculture, Fisheries and Natural Resources NFP Building, Maillard Street Port Louis Fax: +230 212 4427
MEXICO	Dirección General de Normas Secretaría de Comercio y Fomento Industrial Puente de Tecamachalco, N° 6 Lomas de Tecamachalco, Seccion Fuentes, Naucalpan de Juarez, Edo. de Mexico 53950 Mexico D.F. Tel: +52 5 540 26 12/ 520 85 30 Fax: +52 5 72 99 484 Telex: 1775840 IMCEME
MICRONESIA, FEDERATED STATES	Dr. Eliuel K. Pretrick Secretary, Department of Health Services P.O. Box PS70 Palikir, Pohnpei 96941
MONGOLIA	The Director National Centre for Hygiene, Epidemiology and Microbiology Ministry of Health Central Post - PO Box 596 Ulaanbaatar Fax: +976 132 1278
MOROCCO	Division de la Répression des Fraudes Ministère de l'agriculture et de la mise en valeur agricole 25, Avenue des Alaouiyines Rabat Fax: +212 7 763378

MOZAMBIQUE	Codex Contact Point in Mozambique Departamento de Higiene Ambiental Attn. Mr. Evaristo Baquete, Chefe do Departamento Ministerio da Saúde P.O. Box 264 Maputo
MYANMAR	Director Food and Drug Administration Department of Health 35, Min Kyaung Road Yangon 11191 Tel: 71172
NEPAL	Chief Food Research Officer Central Food Research Laboratory Babar Mahal Kathmandu Tel: +977 1 2 14824/ 2 12781
NETHERLANDS	Executive Officer for Codex Alimentarius, Min. of Agriculture, Nature Management & Fisheries Dept. for the Environment, Quality and Health, Room 6302 P.O. Box 20401 2500 EK The Hague Tel: +31 70 3792104 Fax: +31 70 3477552 Email: j.d.m.m.verberne@mkg.agro.nl
NEW ZEALAND	Codex Officer MAF Policy - Ministry of Agriculture and Fisheries P.O. Box 2526 Wellington Tel: +64 4 474 4100 Fax: +64 4 474 4163 Email: RAJ@policy. maf.govt.nz

NICARAGUA	Programa Normalización, Metrología y Control de Calidad (NMCC), Dirección de Tecnología Industrial Del Sandy's Carretera Masaya 1c arriba - Apartado postal N° 8 Managua
NIGER	Division Nutrition Direction Santé Familiale Ministère de la Santé Publique B.P. 623 Niamey
	Tel: +227 72 36 00 Poste 3309 Fax: +227 72 24 24
NIGERIA	Prof. J.A. Abalaka Director General of Standards Standards Organisation of Nigeria Federal Secretariat - Phase 1, 9th fl. P.M.B. No. 2102 (Yaba) Ikoyi, Lagos
	Tel: +234 1 685073/ 682615 Fax: +234 1 681820
NORWAY	Mr. John Race Norwegian Food Control Authority Postboks 8187 0034 Oslo 1
	Tel: +47 2224 6650 Fax: +47 2224 6699 Email: john.race@ snt.dep. telemax.no
OMAN, SULTANATE OF	Director General of Health Affairs Directorate General of Health Affairs Ministry of Health P.O. Box 393 - Darseit Muscat
	Tel: +968 700018 Fax: +968 750562 Telex: 5465

PAKISTAN	The Director-General for Health Ministry of Health, Social Welfare and Population Planning Government of Pakistan Secretariat Block C Islamabad Tel: +92 51 82 09 30 Cable: SEHAT ISLAMABAD
PANAMA	Dirección General de Normas y Tecnología Industrial, Ministerio de Comercio e Industrias Piso 19, Edificio La Lotería Apartado 9658, Zona 4 Panama Tel: +507 227 4749 227 4222 Fax: +507 225 7724
PAPUA NEW GUINEA	Dr. Ian I. Onaga A/Chief Veterinary Officer National Veterinary Laboratory Dept. of Agriculture & Livestock P.O. Box 6372 Boroko NCD Tel: +675 217011/ 217005 Fax: +675 200181/ 214630
PARAGUAY	Dr. Luis Manuel Aguirre Director General, Instituto Nacional de Tecnología y Normalización Avda. General Artigas y General Roa - C.C. 967 Asunción Tel: +595 21 290 160 290 266 Fax: +595 21 290 873

PERU	Dr. Carlos F. Pastor Talledo Director Ejecutivo, Dirección de Higiene Alimentaria y Control de Zoonosis, Dirección General de Salud Ambiental Ministerio de Salud Jr.las Amapolas No.350, Urb. San Eugenio Lima 14 Tel: +51 14 40 2340/ 440 0399 Fax: +51 14 42 6562
PHILIPPINES	Planning and Monitoring Service Department of Agriculture Elliptical Road, Diliman Quezon City, Metro Manila Tel: +63 2 929 8247 Fax: +63 2 928 0590
POLAND	Ministry of Foreign Economic Relations - Quality Inspection Office 32/34 Zurawia str., P.O.Box 25 00-950 Warszawa Tel: +48 22 216421 Fax: +48 22 6214858 Cable: MWGZ-CIS W-wa pl Telex: 813653 pl standard wa
PORTUGAL	Comissão Nacional da FAO, Ministério dos Negócios Estrangeiros Largo do Rilvas 1354 Lisboa CODEX Tel: +351 1 604653 Fax: +351 1 3965161
QATAR	Under-Secretary Ministry of Public Health P.O. Box 3050 Doha Tel: +974 29 20 00 Fax: +974 43 30 19 Telex: 4261

REPUBLIC OF KOREA	Mr. Younghyo HA, Director Technical Cooperation Division International Agriculture Bureau Ministry of Agriculture, Forestry & Fisheries #1, Joongang-Dong Kwachon-si, Kyonggi-do 427-760 Fax: +82 2 507 2095
ROMANIA	Institut Roumain de Normalisation 13, Jean Louis Calderon Code 70201 - Bucarest 2 Tel: +40 1 11 40 43 Fax: +40 1 12 08 23 Telex: 11312 IRS R
RUSSIAN FEDERATION	Institute of Nutrition of the Russian Academy of Medical Sciences - Attention: Drs V.A.Tutelian and A.K. Baturin Ust'Insky Pr., 2/14 109240 Moskva Tel: +7 095 925 1140 Fax: +7 095 230 2812 Telex: 411407
RWANDA	Division-Normalisation et Contrôle de la Qualité Ministère du Commerce et de la Consommation B.P. 476 Kigali Tel: +250 73875/ 73237 Fax: +250 72984 Telex: 502 MINAFFET RW
SAINT KITTS AND NEVIS	Dr. Milton Whittaker Director, Bureau of Standards Ministry of Trade Government of St. Kitts and Nevis Basseterre

SAINT LUCIA	Produce Chemist Ministry of Agriculture, Lands, Fisheries and Cooperatives Government Buildings Castries Fax: +854 453 6314
SAMOA	Chief, Public Health Division Health Department P.O. Box 192 Apia
SAUDI ARABIA	Attention: Public Relations Department (International Relations) Saudi Arabian Standards Organization (SASO) P.O. Box 3437 Riyadh 11471 Tel: +966 1 452 0224 452 0166 Fax: +966 1 452 0167 Telex: SASO SJ 401610
SENEGAL	Comité national du Codex Service de l'alimentation et de la nutrition appliquée au Sénégal (SANAS) Miniastère de la Santé publique Dakar
SEYCHELLES	Director Seychelles Bureau of Standards P.O. Box 648 Victoria (Mahé) Tel: +248 76631 Fax: +248 76151 Telex: 2422 DEPIND SZ
SIERRA LEONE	Mr. A. B. Turay Chief of Standards Central Contact-Codex Alimentarius Com. National Bureau of Standards Ministry of Trade and Industry George Street Freetown

SINGAPORE	Ministry of the Environment Food Control Department Environment Building 40 Scotts Road Singapore 0922 Tel: +65 7319819/ 7327733 Fax: +65 7319843/ 7384468 Telex: RS 34365 ENV
SLOVAK REPUBLIC	Ing. Milan Kováč Výskumný Ústav Potravinársky Food Research Institute Priemyselná 4 - P.O. Box 25 820 06 Bratislava Tel: +42 7 613 55 Fax: +42 7 641 90
SLOVENIA	Dr. Marusa Adamic Hygiene Specialist Institute for Public Health and Social Welfare Trubarjeva 2 61000 Ljubjana Tel: +386 61 32 3245 Fax: +386 61 32 39 55
SOUTH AFRICA	Director: Food Control Department of Health Private Bag X828 0001 Pretoria Tel: +27 12 312 0511 Fax: +27 12 312 0811 Telex: 32-236
SPAIN	Vicesecretaría General Técnica Comisión Interministeriel para la Ordinación Alimentaria (CIOA) Ministerio de Sanidad y Consumo Paseo del Prado 18-20 28071 Madrid Tel: +354 1 596 17 61/ 596 13 46 Fax: +34 1 596 1597/ 596 4409 Telex: 22608/ 44014

SRI LANKA	Director, (Environment and Occupational Health) Ministry of Health - Room 149 385 Deans Road Colombo 10 Tel: +94 1 432050/ 437884 Fax: +94 1 440399
SUDAN	Director-General Sudanese Standards and Metrology Organization (SSMO) P.O. Box 194 Khartoum Tel: +249 11 775247 Fax: +249 11 776359
SURINAME	Ir. G. Hindorie Head - Division of Foreign Relations Ministry of Agriculture, Livestock and Fisheries Cultuurttuinlaan - P.O.Box 1807 Paramaribo Fax: +597 410411
SWAZILAND	The Principal Secretary Att: Director of Health Services Ministry of Health P.O. Box 5 Mbabane Tel: +268 42431 Fax: +268 42092 Telex:2393 MH
SWEDEN	Swedish Codex Contact Point National Food Administration Box 622 S-751 26 Uppsala Tel: +46 18 17 55 00 Fax: +46 18 10 58 48 Email: evlo@msmail. slv.se

SWITZERLAND	Office fédéral de la santé publique Section normes internationales (Attention: Eva Zbinden) 3003 Berne Tel: +41 31 322 95 72 322 95 62 Fax: +41 31 322 95 74 Email: eva.zbinden@ bag.admin.ch
SYRIA	Syrian Arab Organization for Standardization and Metrology P.O. Box 11836 Damascus
TANZANIA	The Tanzania Bureau of Standards P.O. Box 9524 Dar-es-Salaam Tel: +255 51 49041 - 8 Fax: +255 51 48051 Telex: 41667 TBS TZ
THAILAND	The Secretary National Codex Alimentarius Committee of Thailand (TISI) Ministry of Industry Rama VI Street Bangkok 10400 Tel: +66 2 202 3437 Fax: +66 2 247 8741 Telex: 84375 MINIDUS TH (TISI)
THE FORMER YUGOSLAV REPUBLIC OF MACEDONIA	National Institute for Health Protection 6, "5o Udarna Divizija" Str. 91000 Skopje
TOGO	M. le Chargé de liaison du Codex Alimentarius Division de la nutrition et de la technologie alimentaire B.P. 1242 Lomé

TRINIDAD AND TOBAGO	The Chief Chemist and Director of Food and Drugs Chemistry Food and Drugs Division Ministry of Health and Environment 35-37 Sackville Street Port-of-Spain
TUNISIA	M. Ali Ben Gaïd Président Directeur Général Institut national de la normalisation et de la propriété industrielle (INNORPI) B.P. 23 (Cité El Khadhra par rue A. Savary) 1012 Tunis-Belvedere
	Tel: +216 1 785 922 Fax: +216 1 781 563
TURKEY	General Directorate of Protection and Control, Ministry of Agriculture, Forestry & Rural Affairs (Tarim Orman ve Köyisleri Bakanligi, Koruma ve Kontrol Genen Müdürlügü) Akay Cad. N° 3 Bakanliklar Ankara
	Tel: +90 312 1189835 Fax: +90 312 1188005 Telex: 46235 KKGM
UGANDA	The Executive Director Uganda National Bureau of Standards P.O. Box 6329 Kampala
	Tel: +256 41 236606/ 258669
UNITED ARAB EMIRATES	Federal Director Department of Preventive Medicine Ministry of Health P.O. Box 848 Abu Dhabi
	Fax: +917 2 21 27 32/ 31 37 25

UNITED KINGDOM	Head of Branch 'B', Food Labelling and Standards Division Ministry of Agriculture, Fisheries and Food, Room 325 B Ergon House, c/o Nobel House 17 Smith Square London SW1P 3JR
	Tel: +44 171 238 6480 Fax: +44 171 238 6763 Telex: 21271
UNITED STATES OF AMERICA	Executive Officer for Codex U.S. Codex Contact Point Food Safety and Inspection Service US Department of Agriculture Room 311, West End Court Washington D.C. 20250-3700
	Tel: +1 202 418 8852 Fax: +1 202 418 8865 Email: USCODEX@ aol.com
URUGUAY	Ing. Ruperto E. Long Presidente, Laboratorio Tecnológico del Uruguay (LATU) Av. Italia 6201 (C.C. 11500) Montevideo
VANUATU	Mr. Benuel Tarilongi Principal Plant Protection Officer Vanuatu Quarantine Inspection Service, Department of Agriculture and Horticulture Private Mail Bag 040 Port-Vila
	Tel: +678 23130 Fax: +678 24653 Email: qrtnvila@pactok.peg.apc.org
VENEZUELA	Sr. Jefe, Sección de Registro de Alimentos, Ministerio de Sanidad y Asistencia Social Centro Simón Bolivar, Edif. Sur 2 Caracas

VIETNAM	Mr.Nguyen Huu Thien Director-General, Directorate for Standards and Quality 70 Tran Hung Dao Str. Hanoi
	Tel: +84 4 266220 Fax: +84 4 267418
YEMEN	General Director for Measurements Ministry of Economy, Supply and Trade Sana'a
ZAMBIA	Secretary, Food and Drugs Control Ministry of Health P.O. Box 30205 Lusaka
	Fax: +260 1 22 34 35
ZIMBABWE	The Government Analyst The Government Analyst's Laboratory P.O. Box CY 231 Causeway Harare
	Tel: +263 4 792026 Fax: +263 4 708527
NON-MEMBER COUNTRY	
BAHAMAS	H.E. The Prime Minister Office of the Prime Minister P.O. Box 7147 Nassau, N.P.

APPENDIX: GENERAL DECISIONS OF THE COMMISSION

STATEMENTS OF PRINCIPLE CONCERNING THE ROLE OF SCIENCE IN THE CODEX DECISION-MAKING PROCESS AND THE EXTENT TO WHICH OTHER FACTORS ARE TAKEN INTO ACCOUNT[1]

1. The food standards, guidelines and other recommendations of Codex Alimentarius shall be based on the principle of sound scientific analysis and evidence, involving a thorough review of all relevant information, in order that the standards assure the quality and safety of the food supply.

2. When elaborating and deciding upon food standards Codex Alimentarius will have regard, where appropriate, to other legitimate factors relevant for the health protection of consumers and for the promotion of fair practices in food trade.

3. In this regard it is noted that food labelling plays an important role in furthering both of these objectives.

4. When the situation arises that members of Codex agree on the necessary level of protection of public health but hold differing views about other considerations, members may abstain from acceptance of the relevant standard without necessarily preventing the decision by Codex.

---

[1] Decision of the 21st Session of the Commission, 1995.

STATEMENTS OF PRINCIPLE RELATING TO THE ROLE OF FOOD SAFETY RISK ASSESSMENT[1]

1. Health and safety aspects of Codex decisions and recommendations should be based on a risk assessment, as appropriate to the circumstances.

2. Food safety risk assessment should be soundly based on science, should incorporate the four steps of the risk assessment process, and should be documented in a transparent manner.

3. There should be a functional separation of risk assessment and risk management, while recognizing that some interactions are essential for a pragmatic approach.

4. Risk assessments should use available quantitative information to the greatest extent possible and risk characterizations should be presented in a readily understandable and useful form.

---

[1] Decision of the 22nd Session of the Commission

INDEX

—A—

Acceptance of Codex Standards, 4, 18, 19, 23, 24, 27, 29, 30, 31, 32, 33, 35, 36, 37, 38, 39, 40, 57, 65, 86, 105, 106, 107, 109, 110, 147
 Free distribution, 30, 32, 33, 36, 38
 Full acceptance, 30, 31
 Specified Deviations, 23, 27, 30, 31, 36
 Withdrawal or amendment of acceptance, 33
Africa
 FAO/WHO Coordinating Committee, 67, 104
Asia
 FAO/WHO Coordinating Committee, 67, 106

—C—

Cereals, Pulses and Legumes
 Codex Committee on, 68, 101
Cocoa Products and Chocolate
 Codex Committee on, 96
Codex Alimentarius Commission, 4, 5, 6, 7, 8, 9, 10, 11, 12, 13, 14, 15, 16, 17, 18, 19, 20, 21, 22, 23, 24, 25, 26, 27, 28, 33, 34, 36, 43, 44, 48, 49, 50, 51, 52, 53, 55, 67, 69, 71, 73, 78, 79, 85, 86, 88, 89, 90, 91, 95, 98, 101, 102, 103, 104, 105, 106, 107, 108, 109, 110, 111, 113, 147
 Membership, 52, 113
Codex Contact Points, 50, 51, 52, 54, 116
Committees adjourned *sine die*, 27, 28, 96, 97, 99, 100, 101, 102, 104, 111, 112

Contaminants, 27, 29, 37, 38, 42, 65, 67, 70, 71, 75, 76, 77, 78, 81, 87, 88, 93

—E—

Economic Impact Statements, 53, 86
Edible Ices
 Codex Committee on, 101
Elaboration of Codex Standards and Related Texts, 15, 25, 26, 27, 48, 51, 55, 75, 77, 81
Europe
 FAO/WHO Coordinating Committee, 67, 107
Executive Committee of the Codex Alimentarius Commission, 5, 8, 9, 10, 18, 20, 21, 22, 67

—F—

Fats and Oils
 Codex Committee on, 68, 97
Fish and Fishery Products
 Codex Committee on, 100
Food
 Definition, 42
Food additives, 20, 22, 27, 29, 37, 38, 42, 67, 70, 71, 75, 76, 77, 78, 81, 87, 88, 91
Food Additives and Contaminants, 67, 70, 71, 75, 76, 77, 78, 81, 87, 88
 Codex Committee on,, 20, 22, 70, 71, 77, 78, 81, 87, 88
Food Hygiene, 27, 29, 37, 39, 42, 56, 67, 68, 71, 75, 78, 79, 81, 88, 89, 99
 Codex Committee on, 71, 81, 88
Food Labelling, 27, 29, 37, 38, 39, 67, 71, 72, 75, 76, 88, 89, 90, 102, 147
 Codex Committee on, 72, 76, 89
Format of Codex Standards, 29, 38, 69
Fresh Fruits and Vegetables
 Codex Committee on, 102, 103

Fruit Juices
 Joint FAO/ECE Group of Experts, 68, 111

—G—

General Principles
 Codex Committee on, 36, 39, 86
General Principles of the Codex Alimentarius, 23, 24, 25, 27, 36, 37, 39, 54, 56, 67, 77, 78, 79, 86, 111

—I—

Import and Export Inspection and Certification Systems
 Codex Committee on, 68, 75, 94

—L—

Latin America and the Caribbean FAO/WHO Coordinating Committee, 67, 108

—M—

Meat
 Codex Committee on, 98, 99
Meat Hygiene
 Codex Committee on, 99
Membership of the Codex Alimentarius Commission, 113
Methods of Analysis and Sampling, 29, 39, 40, 57, 65, 66, 67, 72, 75, 79, 80, 81, 90, 91
 Analytical Terminology, 58
 Classification of Methods of Analysis, 39, 40, 57, 74, 80
 Codex Committee on, 39, 40, 57, 65, 66, 72, 79, 80, 81, 90
Milk and Milk Products
 Codex Committee on, 85, 104
 FAO/WHO Committee of Government Experts on the Code of Principles Concerning, 85

—N—

Natural Mineral Waters
 Codex Committee on, 68, 104
Nutrition and Foods for Special Dietary Uses, 68, 75, 95
 Codex Committee on, 95

—O—

Observers, 4, 5, 11, 12, 15, 48, 50, 51, 53

—P—

Pesticide Residues, 20, 22, 27, 29, 32, 33, 37, 42, 43, 67, 71, 81, 91, 92, 93
 Codex Committee on, 81, 92
Processed Fruits and Vegetables
 Codex Committee on, 68, 97
Processed Meat and Poultry Products
 Codex Committee on, 68, 99, 111

—Q—

Quick Frozen Foods
 Joint FAO/ECE Group of Experts, 68, 81, 110, 111

—R—

Residues of Veterinary Drugs in Foods, 20, 22, 32, 33, 44, 67, 91, 93
 Codex Committee on, 93
Revision of Codex Standards, 19, 22, 23, 26, 27, 34, 49
Risk Analysis
 Definitions, 44
 Risk Assessment, 45, 147, 148
Rules of Procedure, 5, 7, 8, 9, 10, 11, 12, 13, 14, 15, 16, 17, 48, 53, 85

—S—

Science
 Role of in Codex Decision-Making, 44, 147, 148
Soups and Broths

Codex Committee on, 68, 101
Statutes of the Codex Alimentarius
  Commission, 10, 14, 15, 17, 48
Sugars
  Codex Committee on, 96

—U—

United Nations Economic Commission
  for Europe (UN/ECE), 68, 81, 103,
  110, 111

—V—

Vegetable Proteins
  Codex Committee on, 102

—W—

Weights and Measures, 39, 71